Lab Manual to Accompany
Electronic Communication Systems
SECOND EDITION

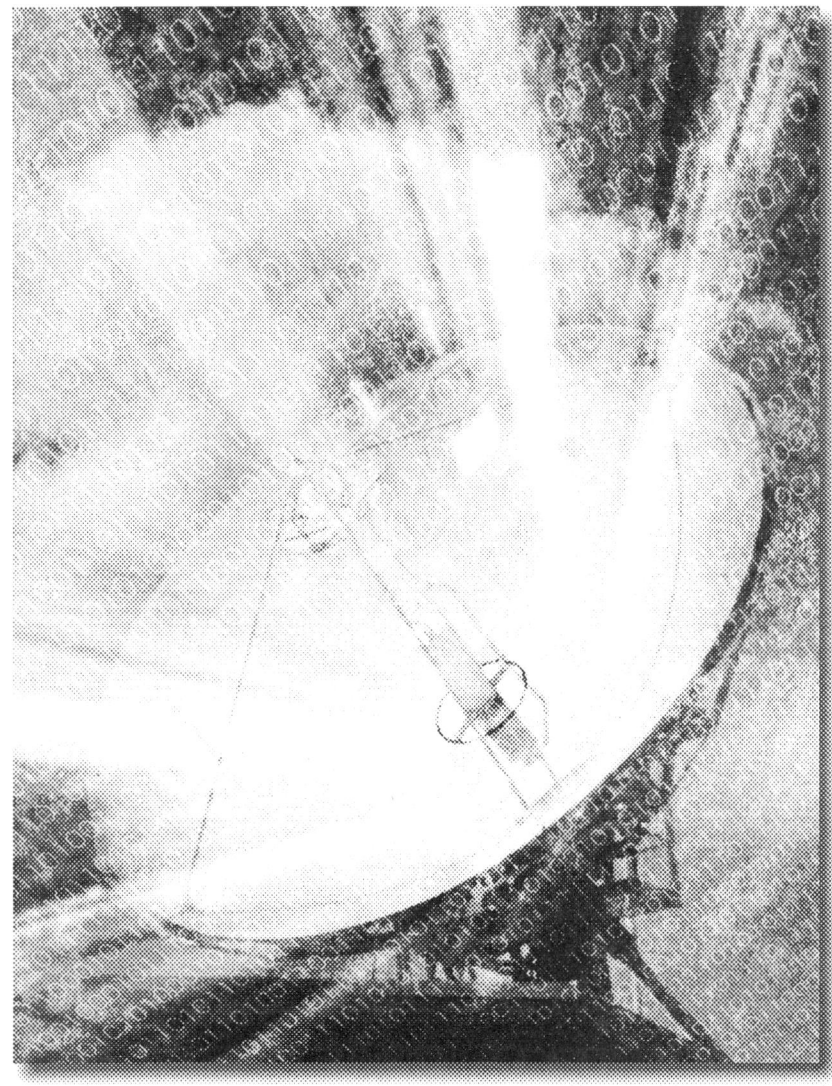

DELMAR

THOMSON LEARNING

Australia • Canada • Mexico • Singapore • Spain • United Kingdom • United States

Lab Manual to Accompany Electronic Communication Systems, 2nd edition

Business Unit Director:
Alar Elken

Executive Editor:
Sandy Clark

Senior Acquisitions Editor:
Gregory L. Clayton

Senior Development Editor:
Michelle Ruelos Cannistraci

Executive Marketing Manager:
Maura Theriault

Channel Manager:
Mary Johnson

Marketing Coordinator:
Karen Smith

Executive Production Manager:
Mary Ellen Black

Production Manager:
Larry Main

Senior Project Editor:
Christopher Chien

Art/Design Coordinator:
David Arsenault

Editorial Assistant:
Jennifer M. Luck

COPYRIGHT © 2002 by Delmar,
a division of Thomson Learning, Inc.
Thomson Learning is a trademark used herein under license

Printed in The United States of America
1 2 3 4 5 XXX 05 04 02 01

For more information contact Delmar,
3 Columbia Circle, PO Box 15015,
Albany, NY 12212-5015.

Or find us on the World Wide Web at
http://www.delmar.com

ALL RIGHTS RESERVED. No part of this work covered by the copyright hereon may be reproduced or used in any form or by any means—graphic, electronic, or mechanical, including photocopying, recording, taping, Web distribution or information storage and retrieval systems—without written permission of the publisher.

For permission to use material from this text or product, contact us by
Tel   (800) 730-2214
Fax   (800) 730-2215
www.thomsonrights.com

Library of Congress Cataloging-in-Publication Data
ISBN 0-7668-4957-0

NOTICE TO THE READER

Publisher does not warrant or guarantee any of the products described herein or perform any independent analysis in connection with any of the product information contained herein. Publisher does not assume, and expressly disclaims, any obligation to obtain and include information other than that provided to it by the manufacturer.

The reader is expressly warned to consider and adopt all safety precautions that might be indicated by the activities herein and to avoid all potential hazards. By following the instructions contained herein, the reader willingly assumes all risks in connection with such instructions.

The Publisher makes no representation or warranties of any kind, including but not limited to, the warranties of fitness for particular purpose or merchantability, nor are any such representations implied with respect to the material set forth herein, and the publisher takes no responsibility with respect to such material. The publisher shall not be liable for any special, consequential, or exemplary damages resulting, in whole or part, from the readers' use of, or reliance upon, this material.

CONTENTS

Lab 1	DECIBELS	1	Lab 11	SSB FUNDAMENTALS	59

Lab 1   DECIBELS   1
Prerequisites / 1
Procedure / 1

Lab 2   RESONANCE   7
Procedure—Series Resonance / 7
Procedure—Parallel Resonance / 10

Lab 3   DISTORTION ANALYZER   13
Prerequisites / 13
Pre-Lab / 13
Procedure / 14

Lab 4   SPECTRUM ANALYZER BASICS   17
Pre-Lab / 17
Procedure / 17

Lab 5   SIGNALS IN THE TIME AND FREQUENCY DOMAIN   21
Prerequisite / 21
Pre-Lab / 21
Procedure / 22

Lab 6   MIXERS   27
Frequency Multipliers / 27
Frequency Conversion and Mixer Action / 29

Lab 7   NOISE   33
Procedure / 33

Lab 8   AMPLITUDE MODULATION WAVEFORM MEASUREMENTS   39
Procedure / 39

Lab 9   RECEIVER SENSITIVITY   47
Procedure / 47

Lab 10   BALANCED MODULATORS   53
Pre-Lab / 53
Procedure / 55

Lab 11   SSB FUNDAMENTALS   59
Procedure / 59

Lab 12   RECEIVER SELECTIVITY   65
Introduction / 65
Procedure / 66

Lab 13   THE PHASE LOCKED LOOP   69
Introduction / 69
Pre-Lab / 70
Procedure / 71

Lab 14   FM MODULATION AND DEMODULATION   75
Frequency Demodulation Using a Phase Locked Loop (PLL) / 75

Lab 15   FM RECEIVER SENSITIVITY   81
FM Receiver Sensitivity Measurements / 81
SINAD Sensitivity / 84
Squelch Control / 85

Lab 16   FM CAPTURE EFFECT   87
Introduction / 87
Procedure / 88

Lab 17   DATA TERMINAL I/O TROUBLESHOOTING   93
Prerequisites / 93

Lab 18   ASYNCHRONOUS COMMUNICATIONS   99
Prerequisite / 99
Procedure / 100

Lab 19   RS-232 SIGNALS   105
Pre-Lab / 105
Introduction / 106

Lab 20   UARTS   111
Prerequisites / 111
Procedure / 111

Lab 21 MODEMS AND FILE
TRANSFER PROTOCOLS 121
  Prerequisites / 121
  Procedure / 121

Lab 22 PULSE CODE MODULATION
AND CODECS 133
  Purpose / 133
  Introduction / 133

Lab 23 DIRECTIONAL COUPLERS
AND VSWR 145
  Prerequisites / 145
  Introduction / 145
  Procedure / 146

Lab 24 TIME DOMAIN
REFLECTOMETRY (TDR) 151
  Introduction / 151
  Procedure / 151

Lab 25 SATELLITE LINK
MEASUREMENTS 155
  Prerequisites / 155
  Introduction / 155
  Pre-Lab / 156
  Procedure / 158

Lab 26 SYSTEM RISE TIME AND
DIGITAL TRANSMISSION 167
  Prerequisites / 167
  Procedure / 167

Appendix A
MANUFACTURER'S DATA SHEETS
FOR LAB 10 171

Appendix B
MANUFACTURER'S DATA SHEETS
FOR LAB 20 175

DECIBELS

LAB 1

Name: _____ Date: _____

OBJECTIVES:

Upon completion of this lab, you will be able to:
- Use an ac voltmeter to establish a desired dBm level
- Measure the power level in dBm using an ac voltmeter
- Measure the attenuation in dB of a resistive pad

TEST EQUIPMENT:

- AC voltmeter (preferably with dB scale)
- Oscilloscope
- Function generator
- 50 ohm termination; miscellaneous resistors

Prerequisites

To perform this lab, the student should have received instruction on the following topics:
- the deciBel (dB) as a measure of gain and loss
- the reference power levels dBm and dBW

Procedure

Figure 1-1 is a conceptual diagram of the equipment setup. Notice that both the oscilloscope and the voltmeter are measuring the voltage across a 50 Ω load. (Both the voltmeter and oscilloscope have very high input impedances.)

Adjust the output of the function generator to produce a 2.24 rms volt 1 kHz sine wave. Use the voltmeter to make this setting.

If the voltmeter reading across the 50 ohm load is 2.24 volts rms, the power dissipated in the 50 Ω load should be +20 dBm.

The function generator is therefore set to output +20 dBm.

1. Show by calculations that 2.24 volts rms across a 50 ohm load resistor dissipates +20 dBm.

If the voltmeter you are using has a dB scale, make a check of the dB reading on the dB scale. Unless the voltmeter is specifically calibrated for measuring dBm across a 50 ohm load, the dB reading you see will *not* be +20. (A correction factor can be used to correct the reading, but the point is to be aware that dBm readings on a voltmeter can be misleading to the unwary.)

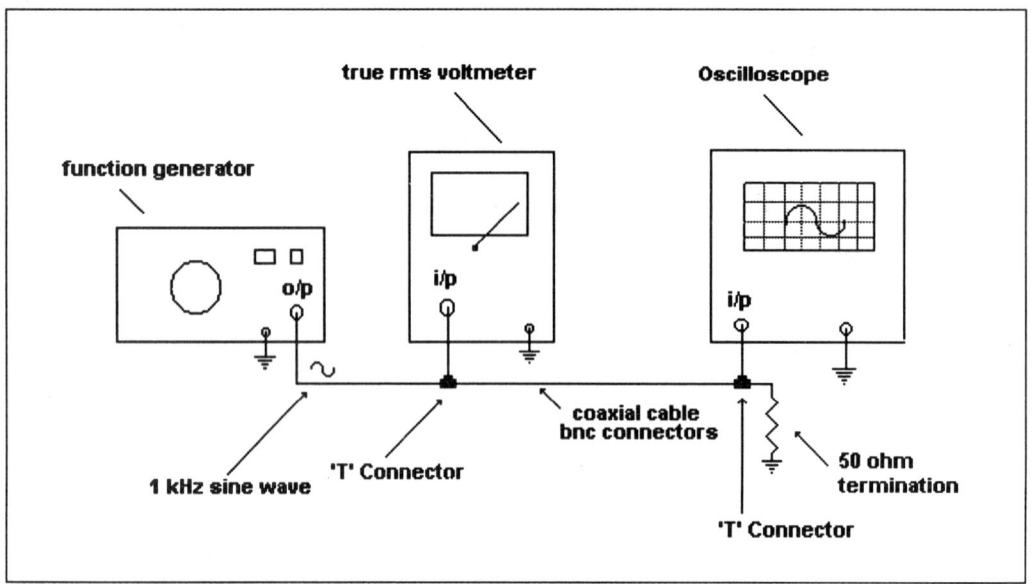

FIGURE 1-1

2. If the voltmeter has a dB scale, what is the dB reading?

Monitor the 1 kHz sine wave with the oscilloscope and measure the peak-to-peak voltage of the sine wave across the 50 ohm load.

3. What would be the theoretical V_{p-p} in volts for the oscilloscope reading? Show the calculation details.

4. What is the measured V_{p-p}?

Remove the 50 ohm termination and note that both the reading on the oscilloscope and the voltmeter have changed.

5. What is the new V_{p-p} indication on the oscilloscope?

6. What is the new voltmeter reading in volts?

7. If the voltmeter has a dB scale, what is the new reading in dB and what is the change in dB from the result in question 2?

8. Explain the cause of the change in the readings. (Hint: Use the voltage divider rule and remember that both the scope and voltmeter have very high input impedances.)

9. From the results obtained above, calculate the value of the output impedance of the function generator.

If the output impedance of the generator is 50 ohms, the voltage readings should have doubled when the 50 ohm termination was removed from the circuit. If the voltage doubled, the dB reading should have increased by +6 dB since the voltmeter's dB scale is calibrated to 20 log (voltage).

Without changing the output setting of the function generator, construct the resistive pad shown in Figure 1-2.

The resistors R_1, R_2, and R_3 are the components for a resistive pad which has been designed to provide 20 dB of attenuation while matching a 600 ohm resistive load to 50 ohms. (Use series and parallel combinations of resistors to achieve the resistor values called for in the diagram.)

Use the voltmeter to measure the output voltage of the function generator when connected to the resistive pad, as shown conceptually in Figure 1-2.

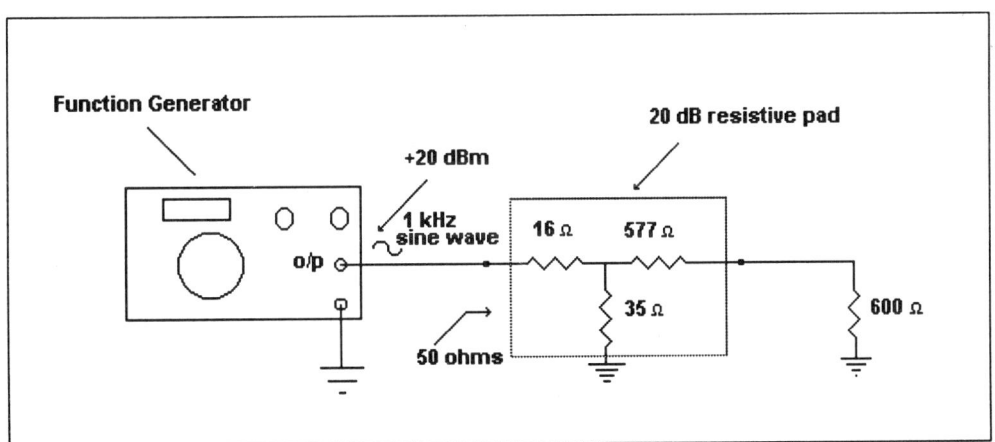

FIGURE 1-2

10. What is the rms voltage at the generator output? What is the output power in dBm from the generator?

The output of the function generator was previously set for an output of +20 dBm or an rms voltage of 2.24 volts across a 50 ohm termination. Since the input resistance of the resistive pad is designed to be 50 ohms when matched to 600 ohms, there should be very little change in the voltage reading or the dB scale reading on the voltmeter. Any change will be due to tolerances in the resistors used to construct the resistive pad.

Using the voltmeter, measure the voltage across the 600 ohm resistor at the output of the 20 dB resistive pad.

11. What is the rms voltage across the 600 ohm resistor?

Using the rms voltage reading measured across the 600 ohm resistor, you should be able to calculate the power level in dBm at this point in the circuit.

12. Calculate the power level in dBm at the pad output.

Since the resistive pad has about 20 dB of attenuation and about 20 dBm was the input power to the pad, the output power from the pad should be close to 0 dBm. This follows from the following relationship:

(Output power in dBm) = (Input power in dBm) − (20 dB of attenuation)

(0 dBm) = (+20 dBm) − (20 dB) = (0 dBm)

The following relationship allows the calculation of power loss or gain for a circuit when only the voltage gain is measured.

$$A_P \text{ (dB)} = A_V \text{ (dB)} + 10 \log \frac{Z_i}{Z_o}$$

where

$$A_V \text{ (dB)} = 20 \log (V_{out} / V_{in})$$

and

Z_i = 50 ohms

Z_o = 600 ohms

13. Use this equation and the measured values of voltage at the input and output of the resistive pad to calculate the dB power loss in the pad.

 Answer: You should have calculated a 20 dB loss.

RESONANCE

LAB 2

Name: _____ Date: _____

OBJECTIVES:

Upon completion of this lab, you will be able to:

- Measure the response of a series and parallel resonant circuit
- Measure the resonant frequency, bandwidth, and Quality factor of a series and parallel resonant circuit

TEST EQUIPMENT:

- Dual channel oscilloscope
- Function generator
- DMM
- Miscellaneous parts — Resistors 330 Ω, 10 kΩ
  — Capacitor 1.0 nF
  — Inductor 2.4 mH

Procedure—Series Resonance

You will be constructing the circuit shown conceptually in Figure 2-1, but first you must make some preliminary measurements and calculations.

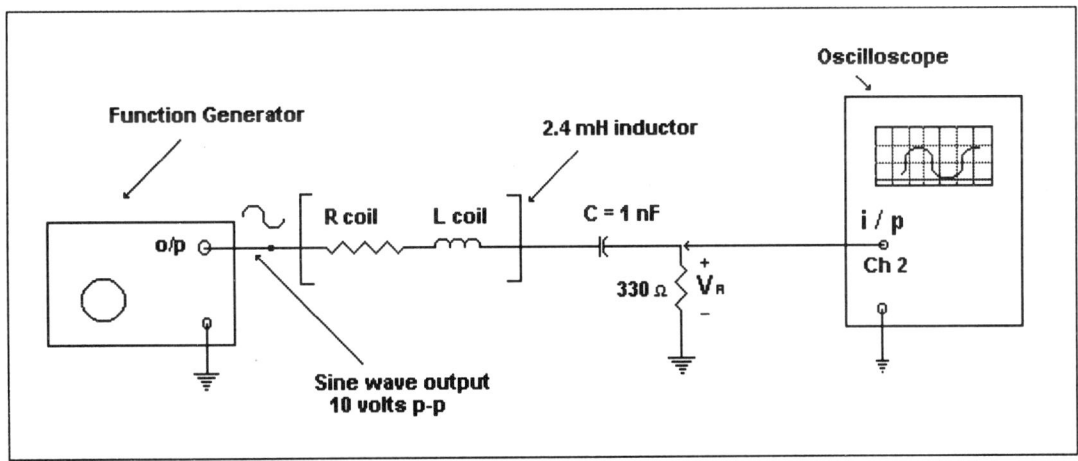

FIGURE 2-1

Using the values for the inductor and capacitor shown in Figure 2-1, calculate the resonant frequency for the circuit. Using the ohmeter function on the DMM, measure the resistance of the inductor.

1. What was the measured resistance of the inductor?

2. What is the calculated value for the resonant frequency in Figure 2-1?

3. If the inductor's resistance and the 330 ohm load resistor make up the total series resistance of this resonant circuit, what is the circuit Quality factor (Q_{CCT}).

Connect the circuit as shown conceptually in Figure 2-1 and use the channel 1 probe of the oscilloscope to monitor the input of the circuit.

Adjust the function generator for a 10 volt p-p sine wave and maintain this voltage at the input to the circuit at all frequency settings.

Connect the channel 2 probe of the oscilloscope to monitor the voltage across the 330 Ω resistor.

Vary the frequency setting of the function generator output and notice that as you do so the voltage (V_R) across the 330 Ω resistor varies and goes through a maximum voltage. The frequency which gives the maximum voltage across the resistor is, of course, the resonant frequency of the circuit. You should notice that at this frequency the sine waves at the input of the circuit (channel 1) and across the 330 Ω resistor (channel 2) are in phase.

4. At what measured frequency did your circuit go into resonance?

The measured resonant frequency should be the same or relatively close to the theoretical value you calculated.

Complete Table 2-1 to investigate how the resonant response varies with frequency. (Remember to keep the input voltage at 10 volts p-p at all frequencies.)

Table 2-1	
Frequency	Voltage across 330 Ω resistor (V_R)
5 kHz	
10 kHz	
20 kHz	
30 kHz	
100 kHz	
200 kHz	
400 kHz	
800 kHz	
1.5 MHz	
2.0 MHz	

Using log-log graph paper, plot the response of V_R versus frequency.

Add extra measurement points to those suggested by Table 2-1 if it helps define your graph more clearly.

An inspection of your graph of V_R versus frequency should show two frequencies on either side of resonance where the voltage across the 330 Ω resistor has declined to 0.707 of the maximum voltage measured at resonance. These are called the upper and lower half power frequencies and the difference between them defines the bandwidth of the circuit.

5. What is the measured value of the upper half power frequency?

6. What is the measured value of the lower half power frequency?

7. What is the bandwidth of the resonant circuit?

The resonant frequency divided by the Quality factor (Q_{CCT}) of a circuit is the circuit bandwidth. Clearly the Quality factor is an important parameter in determining the behaviour of a resonant circuit. Using the measured value for the resonant frequency and the measured bandwidth, you can arrive at an estimate of the Quality factor for the circuit.

8. What is the Quality factor for this resonant circuit?

One of the important and useful characteristics of a series resonant circuit is the resonant rise in voltage across both the capacitor and the inductor at the resonant frequency. At resonance, the magnitude of the voltage across the capacitor will be the Q_{CCT} times the input voltage, and so will be the voltage across the inductive reactance component of the inductor. In this lab, the function generator has been set to input 10 volts p-p to the resonant circuit. If the Q_{CCT} had a value of 3, the voltage across the capacitor would be 30 volts p-p.

Adjust the frequency of the function generator so that the circuit is again at resonance. (Remember that at resonance V_R is in phase with the input 10 volt p-p sine wave.)

Interchange the 330 Ω resistor with the capacitor and measure the peak-to-peak voltage across the capacitor. Now interchange the inductor with the capacitor and measure the peak-to-peak voltage across the inductor. Use channel 2 of the oscilloscope to make these measurements.

Remember in both cases to check that the voltage at the input to the circuit is still 10 volts p-p on channel 1 of the oscilloscope.

9. What was the measured peak-to-peak voltage across the capacitor? Across the inductor?

10. Calculate the Q_{CCT} from the magnitude of the voltage measured across the capacitor.

11. How does this value compare with the Q_{CCT} measured in question 8?

12. How well did the theoretical value for the Q_{CCT} compare with the measured values?

Procedure—Parallel Resonance

You will be constructing the circuit shown conceptually in Figure 2-2, but first you must make some preliminary calculations for this circuit as outlined in Table 2-2.

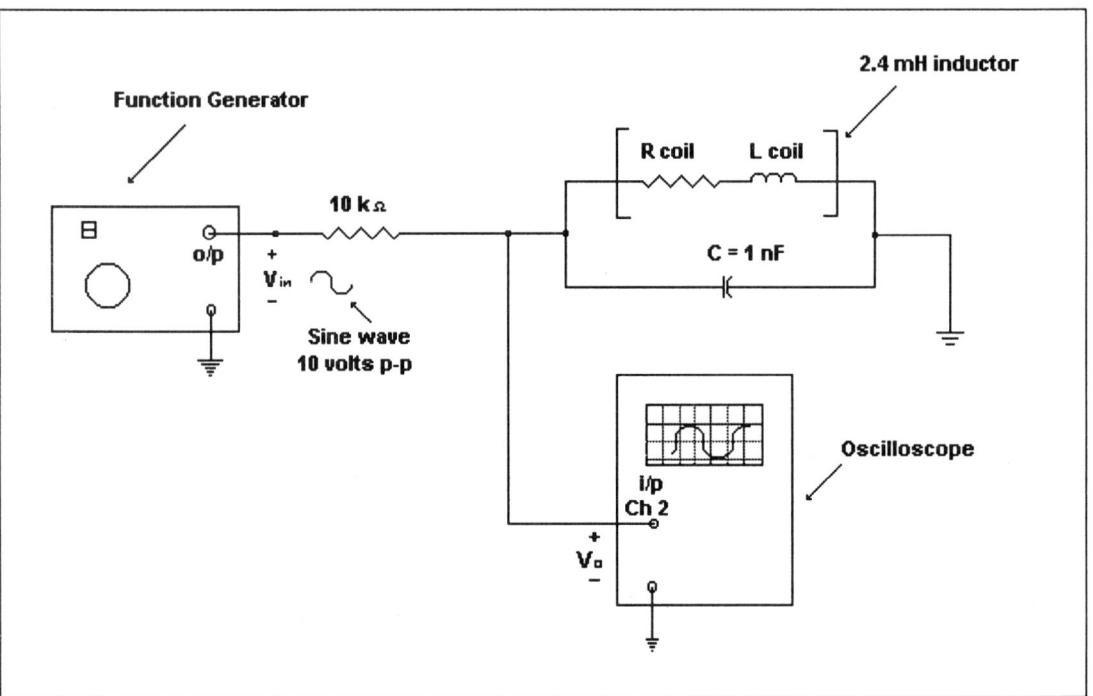

FIGURE 2-2

Table 2-2		
Circuit parameter		Calculated value
Inductor resistance		(as measured for series resonance)
Resonant frequency	f_0	
Quality factor of circuit (include effects of 10 kΩ resistor and coil losses)	Q_{cct}	
Bandwidth	BW	
Half power frequencies	f_1 f_2	

Adjust the function generator to 10 volts p-p at 100 kHz.

Construct the circuit shown conceptually in Figure 2-2 and connect the function generator to the input of the circuit.

Use channel 1 of the oscilloscope to monitor the input voltage V_{in} to the kohm resistor and adjust this voltage to be 10 volts p-p. At each frequency setting, maintain the input voltage V_{in} to be a constant 10 volts p-p.

Monitor the output voltage V_o with channel 2 of the oscilloscope.

Using the frequencies suggested in Table 2-2, measure the output voltage V_o and enter the data in Table 2-3. Use the data and plot the response of the circuit (V_o versus frequency) using log-log graph paper. The circuit is at resonance when the phase difference between V_{in} and V_o is zero degrees. Indicate the resonant frequency on your graph.

Table 2-3	
Frequency	V_o (volts p-p)
5 kHz	
10 kHz	
20 kHz	
30 kHz	
40 kHz	
50 kHz	
60 kHz	
80 kHz	
100 kHz	
120 kHz	
150 kHz	
300 kHz	
600 kHz	
1 MHz	
2 MHz	

Answer the following questions by analyzing the data gathered in doing this lab. Where a calculation is involved in arriving at your answer, show your work.

13. What is the measured resonant frequency and how well did it compare with the theoretical value?

14. When $V_o = 0.707\ V_{o\ max}$, what are the two half power frequencies f_1 and f_2?

15. What is the measured bandwidth of this circuit?

16. What is the measured Quality factor (Q_{cct}) for this circuit?

17. How well did your results compare with theory?

DISTORTION ANALYZER

LAB 3

Name: _____ Date: _____

OBJECTIVES:

Upon completion of this lab, you will be able to:

- Use Fourier series and the definition for total harmonic distortion (THD) to calculate the distortion of a signal
- Use a distortion analyzer to measure the THD of a signal

TEST EQUIPMENT:

- Distortion analyzer*
- Function generator
- Oscilloscope

---
*Amber 3500 or any general purpose distortion analyzer

Prerequisites

To perform this lab, the student should have received instruction on the following topics:

- Fourier series representation of standard waveforms such as triangle and square waves
- The basic definition of total harmonic distortion (THD) and the ability to calculate the THD of a waveform or signal

Pre-Lab

The normal use of a distortion analyzer is to measure the distortion in a signal such as a sine wave where the expected % distortion is very small. This is what we will do in this lab, except that we will first make some measurements of distortion on waveforms of known distortion such as the triangle and square waveforms. Of course, these waveforms are not sine waves, but we can use Fourier series to calculate their harmonic content and then measure the THD using the distortion analyzer. If we do our calculations correctly, we should measure THDs close to what we calculate and thus gain insight and confidence in our measurement technique.

14   LAB 3 DISTORTION ANALYZER

1. Using a table of Fourier series, write the Fourier series of a 1 V peak 1000 Hz square wave. Using the coefficient of 1000 Hz as fundamental, calculate the THD using the first five harmonics and the THD mathematical formula. Express your answer as a %.

$$THD = \frac{\sqrt{\sum_{1}^{n} A_i^2}}{A_0}$$

 Answer: You should have calculated a THD of 41.4%.

2. Repeat the calculation in (1) for a 1 V peak 1000 Hz triangle wave.

 Answer: You should have calculated 12%.

Procedure

The equipment hookup will be conceptually shown in Figure 3-1. Your instructor will provide a detailed hookup diagram for the specific equipment you are using. Use the instructor's diagram and connect the equipment, but before doing so **make sure no power is turned on**. When you are finished connecting the equipment, have the instructor check your setup.

 Instructor's signature: _____

When you have the instructor's checkoff, turn on the power to the equipment.

Adjust the output of the function generator for a 1 V peak 1000 Hz square wave and measure the % distortion with the analyzer. (Instructions on how to use the analyzer for this purpose are attached to the back of this lab or will be provided to you by your instructor.)

3. What was the measured distortion of the square wave?

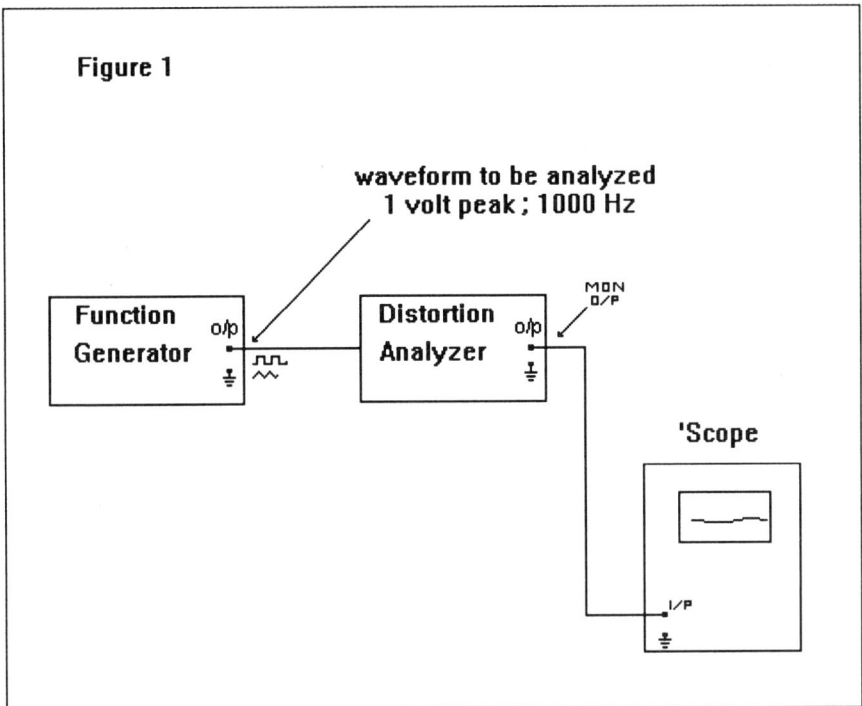

FIGURE 3-1

4. How did the measured value compare with theory?

While measuring the distortion, you should observe the distortion analyzer output signal that is connected to the oscilloscope. What you are observing is the square wave with the 1000 Hz fundamental filtered out.

5. Sketch the observed output seen on the oscilloscope.

Now adjust the output of the function generator for a 1 V peak 1000 Hz triangle wave and measure the % distortion with the analyzer.

6. What was the measured distortion of the triangle wave?

7. How did the measured value compare with theory?

Now change the function generator to a 1 V peak 1000 Hz sine wave. Measure the distortion in %.

8. What is the % distortion of the generator sine wave?

9. Observe and sketch the distortion as viewed on the scope.

Set the function generator to produce a sine wave with a voltage of 1 V p-p as measured on the oscilloscope. Then use the distortion analyzer in its voltmeter function to measure the same voltage. Repeat this procedure with a square wave of the same p-p voltage.

10. What was the reading on the analyzer for the sine wave?

11. What was the reading on the analyzer for the square wave?

12. Were they what you would have expected? Explain. (Hint: If the square wave reading does not seem right, check to see if the analyzer meter is average responding or true rms.)

SPECTRUM ANALYZER BASICS

LAB 4

Name: _____ Date: _____

OBJECTIVES:

Upon completion of this lab, you will be able to use a spectrum analyzer to:

- Measure the amplitude and frequency of the calibration signal
- Resolve and measure the spectral content of a signal

TEST EQUIPMENT:

- Spectrum analyzer
- Miscellaneous signal sources —calibration signal on analyzer
  —AM signal

> *NOTE TO THE INSTRUCTOR:* Connecting the video output from the spectrum analyzer to a large screen video monitor allows the demonstration of this material to a classroom. This would be a useful option if the spectrum analyzer is a scarce resource.

Pre-Lab

Before starting the lab, it would be useful to review some spectrum analyzer basics by referring to your text and completing the following exercise.

1. Draw and label the block diagram of a swept frequency spectrum analyzer. Also indicate on your diagram which block controls each of the four main functions of **span**, **resolution bandwidth**, **reference level**, and **center or start frequency**.

Procedure

For the following portions of the lab, your instructor will provide a diagram of the front panel of the spectrum analyzer identifying key controls as well as access to the analyzer's user manual.

17

NOTE: A spectrum analyzer is easily damaged if the input signal is too large. Do not connect any signal source to the analyzer other than specified and until told to do so by the instructor. If you are unsure about what you are doing, ask the instructor.

At this point in the lab there should be no signal connected to the input of the analyzer and **power should be off.** Check that this is so.

Have the instructor check your work station before proceeding.

Instructor's signature: _____

Turn on the power to the spectrum analyzer. The analyzer screen should display information on the spectrum analyzer's settings. (On older analyzers this information has to be obtained from the knob settings.)

2. From the analyzer screen display, find the following information.

   Reference level in dBm = _____

   Vertical scale factor in dB/ = _____

   Center frequency in MHz = _____

   Span in MHz/ = _____

   Resolution bandwidth in MHz = _____

3. On the screen below print or indicate where you found each of the readings.

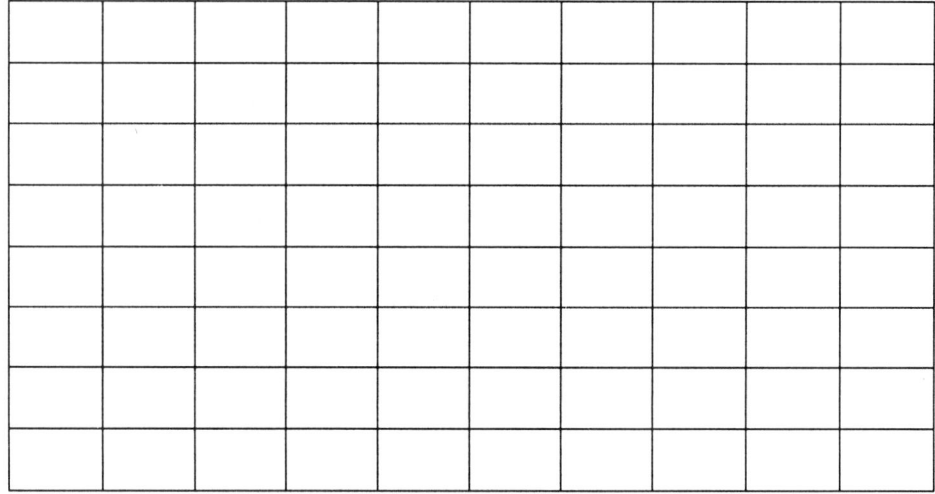

Have your instructor check your conclusions and readings.

 Instructor's signature: _____

Let us begin using the spectrum analyzer by making a simple measurement of a signal. Since the spectrum analyzer has a calibration signal available, we will use this signal. Using the diagram of the front panel of the analyzer provided by the instructor, locate on the front panel the calibration signal output and the location for signal input. Have the instructor check that you are correct before proceeding.

Instructor's signature: _____

The instructor will connect a 50 Ω coaxial cable between the calibration signal output and the analyzer signal input. The calibration signal spectra should appear on the analyzer display.

Find printed on the analyzer front panel the answers for the following questions.

4. What is the frequency in MHz of the calibration signal?

5. What is the amplitude in dBm of the calibration signal?

In order to manipulate the display of this calibration signal we will use the three basic controls of span, center frequency, and reference level. Have your instructor give you a demonstration showing the location of these controls on the analyzer front panel and a quick introduction on how to use them.

Demonstrate in the instructor's presence that you can use these controls to change the display of the calibration signal. Note the effect of varying the span, center frequency, and reference level controls.

 Instructor's signature: _____

6. On the display provided, draw what the screen looks like. Indicate reference level, center frequency, span, and scale factor.

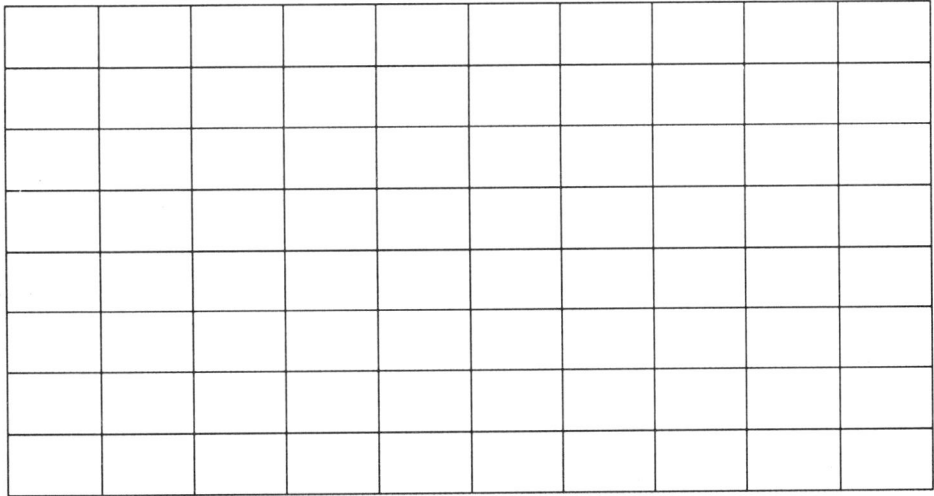

The instructor will provide through a coaxial cable from his workstation an amplitude modulated signal for you to view on the spectrum analyzer. The AM signal will have a carrier strength of −10 dBm and a carrier frequency of 25 MHz. The AM signal will have two side frequencies, both equally spaced above and below the 25 MHz carrier and both with the same signal strength.

7. Use the spectrum analyzer to determine the amplitude in dBm and frequency in MHz of the three components in the AM signal. Sketch your measured results on the display below. Indicate the amplitude and frequency of each signal.

SIGNALS IN THE TIME AND FREQUENCY DOMAIN

LAB 5

Name: _____ Date: _____

OBJECTIVES:

Upon completion of this lab, you will be able to:
- Predict the spectral content of a square or triangle waveform from the use of Fourier series
- Measure and verify the spectral content of a waveform using a spectrum analyzer
- Measure the spectral content of a signal after filtering

TEST EQUIPMENT:

- Oscilloscope
- Function generator
- Spectrum analyzer
- Materials — resistors 1 kΩ, 4.7 kΩ, 10 kΩ, 51 Ω
  — capacitors 0.1 µF, 1.5 nF, 330 pF
- Optional: Video monitor connected to video output of the spectrum analyzer for class demonstration

Prerequisite

To perform this lab, the student should have received instruction on Fourier series representation of standard waveforms such as triangle and square waves.

Pre-Lab

1. Using a table of Fourier series, write the Fourier series of a 1 V peak-to-peak 100 kHz square wave.

2. Repeat the calculation in (1) for a 1 V peak-to-peak 100 kHz triangle wave.

Procedure

Figure 5-1 is a conceptual diagram of the equipment setup. Assemble the equipment needed, but do NOT turn on the power to any equipment. Do NOT connect the spectrum analyzer into the circuit.

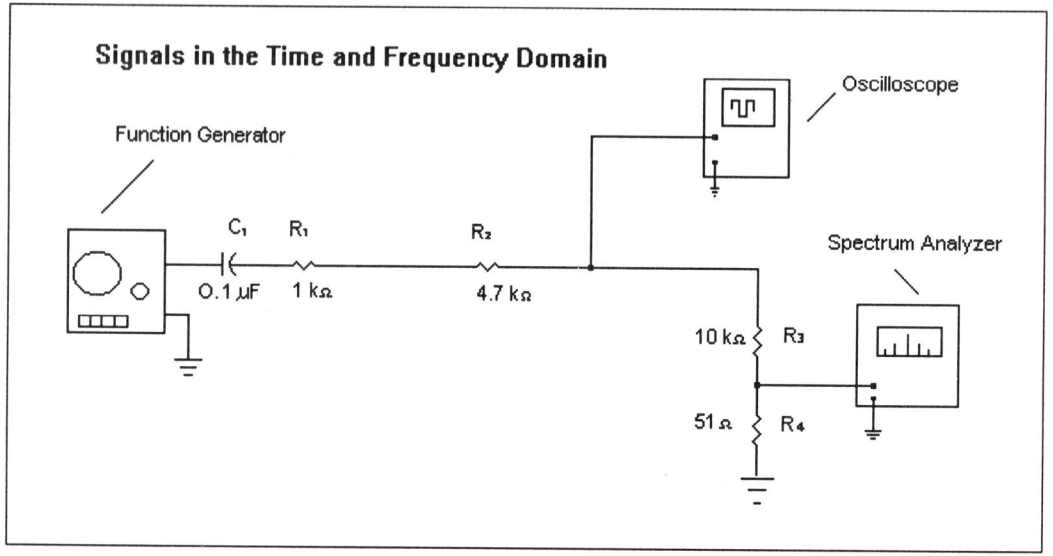

FIGURE 5-1

C1 is in the circuit to make sure there is no dc offset from the function generator which could possibly damage the spectrum analyzer. The resistor chain forms an attenuator to protect the spectrum analyzer from too large an ac signal.

3. How much attenuation in dB does the resistor network provide for the spectrum analyzer? Show your calculations. Assume the input impedance for the analyzer is 50 ohms.

Turn on the power to the function generator and the scope alone. (At this time the spectrum analyzer is still NOT connected to the circuit.) Set the generator to produce a sine wave with a peak-to-peak voltage of 1 V at a frequency of 100 kHz, with no dc offset. Observe the signal with the oscilloscope. Ask the instructor to check your setup and assist with connecting the spectrum analyzer into the circuit.

 Instructor's signature: _____

Now turn on the power to the analyzer.

4. Sketch both the time and frequency displays shown on the oscilloscope and spectrum analyzer. Does the sine wave have any harmonics?

   (a) Oscilloscope

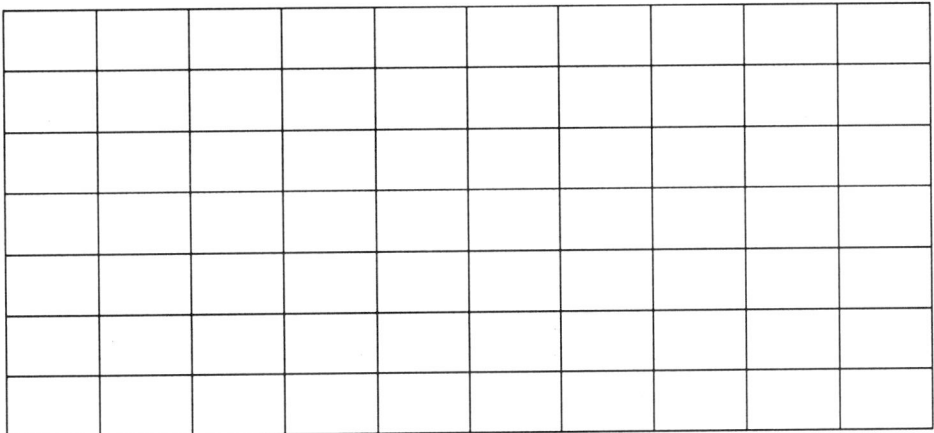

   (b) Spectrum Analyzer

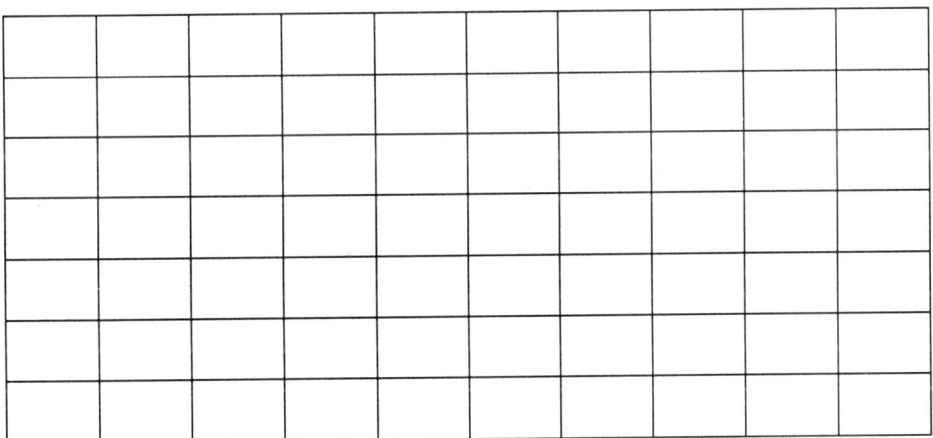

Change the signal from the function generator to a 1 V p-p square wave. Adjust the settings on the spectrum analyzer to display in the linear mode. This will make it easier for us to compare the results we obtained with our previous calculations in question 1.

5. Sketch both the time and frequency displays obtained with the oscilloscope and the spectrum analyzer indicating values. Compare the spectrum to the values predicted by Fourier series. Is the third harmonic 1/3 the amplitude of the fundamental, and the fifth harmonic 1/5 the fundamental amplitude as theory predicts?

(a) Oscilloscope

(b) Spectrum Analyzer

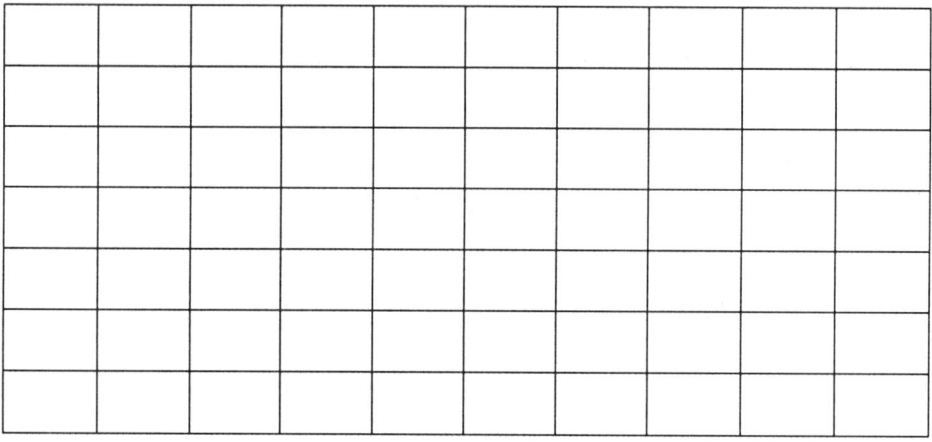

Change the function generator to output a 1 V p-p triangle waveform.

6. Sketch the oscilloscope and spectrum analyzer displays. How do the results compare with the values calculated from theory?

(a) Oscilloscope

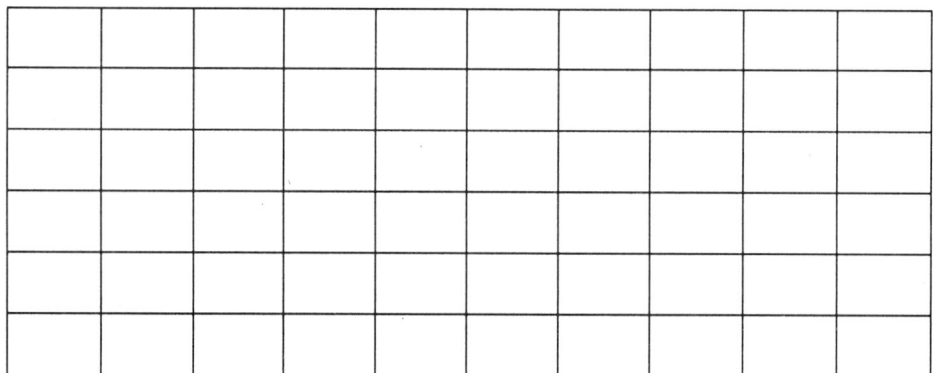

(b) Spectrum Analyzer

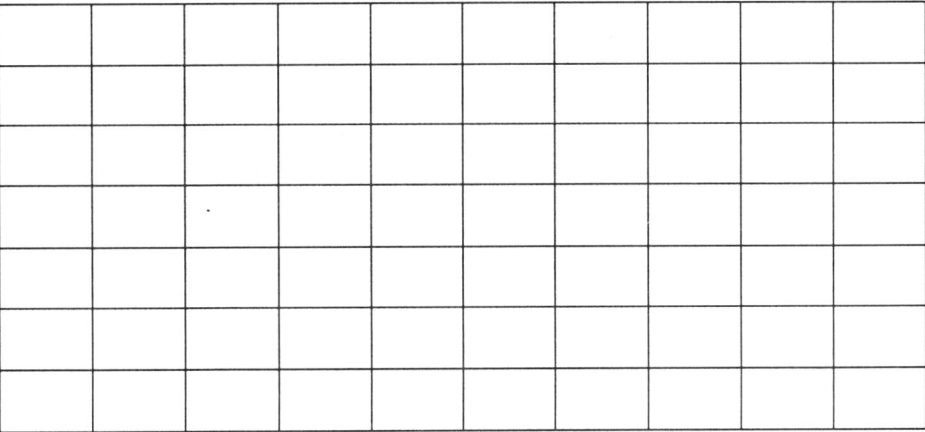

Add two capacitors to the circuit as shown conceptually in Figure 5-2. To avoid accidently causing a damaging voltage on the input of the spectrum analyzer, disconnect the analyzer from the circuit while adding the capacitors. Adding the capacitors to the circuit will create a second order lowpass filter with a cutoff frequency of about 100 kHz.

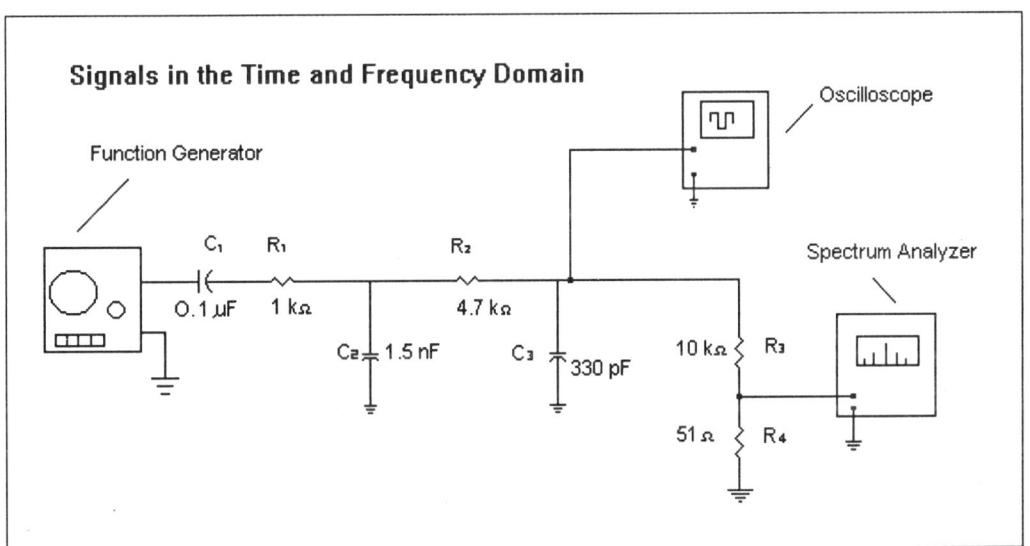

FIGURE 5-2 Capacitors added to provide filtering

Observe the triangle wave, square wave, and sine wave again on both the oscilloscope and spectrum analyzer.

7. How do the results compare with theory? Can you explain what you observe for the

   Triangle wave

   Square wave

   Sine wave

The addition of the capacitors introduced filtering into the circuit. The filter is as shown in Figure 5-3 below.

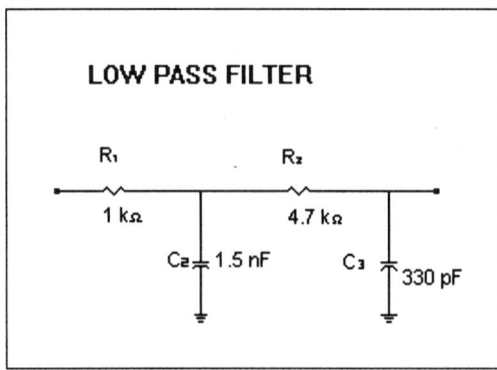

FIGURE 5-3

8. For the filter circuit shown in Figure 5-3, calculate the output voltage at 1000 Hz, 100,000 Hz, and again at 200,000 Hz if the input input voltage is 1 V p-p at each frequency.

9. Optional: Derive the transfer function of the filter and sketch the Bode plot of the filter response (dB vs log f) on semi-log graph paper. Indicate clearly the cutoff or break frequency.

MIXERS

LAB 6

Name: _____   Date: _____

OBJECTIVES:

Upon completion of this lab, you will be able to:
- Demonstrate frequency multiplication using a signal diode and a filter
- Demonstrate frequency conversion using a signal diode and a filter

TEST EQUIPMENT:

- Two function generators
- Dual channel oscilloscope
- Signal diode   (1N914)
- Capacitor   1 nF
- Inductor   2.4 mH
- Resistors   10 kohm (3)

Frequency Multipliers

Introduction

A sine wave signal can be doubled in frequency by driving a non-linear circuit with the signal whose frequency is to be doubled and then filtering the resulting distorted signal at the frequency of the second harmonic. A frequency tripler is achieved by filtering at the frequency of the third harmonic.

Procedure

In this lab, you will use a signal diode for the non-linear circuit. Adjust the output of the function generator for a 3 V p-p sine wave at 50 kHz. Refer to Figure 6-1 for a conceptual diagram of the circuit setup.

Observe the output signal across the 10 kΩ load resistor using channel 2 of the oscilloscope and note the severe distortion. Compare the distorted output with the relatively clean input sine wave on channel 1.

1. Is there a dc voltage component in the distorted output signal?

The severe distortion of the sine wave is due to the harmonic content, and the dc component is there due to the rectifier action of the diode.

Modify your circuit so that the distorted signal is fed to a filter as shown conceptually in Figure 6-2.

As can be seen from Figure 6-2, the filter is a 2.4 mH inductor in parallel with a 1 nF capacitor. A parallel connection of a capacitor and an inductor can go into resonance providing filtering at the parallel resonant frequency.

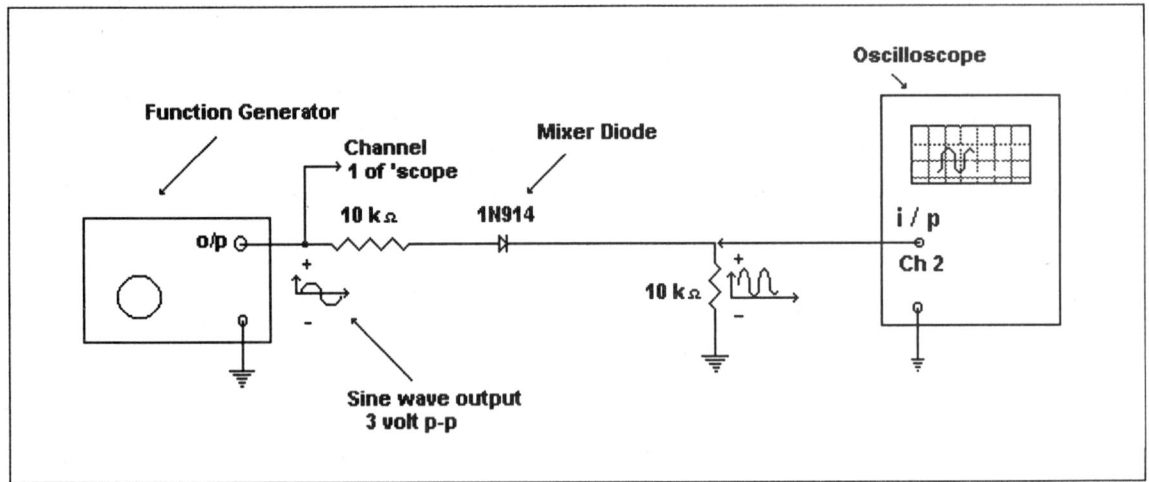

FIGURE 6-1

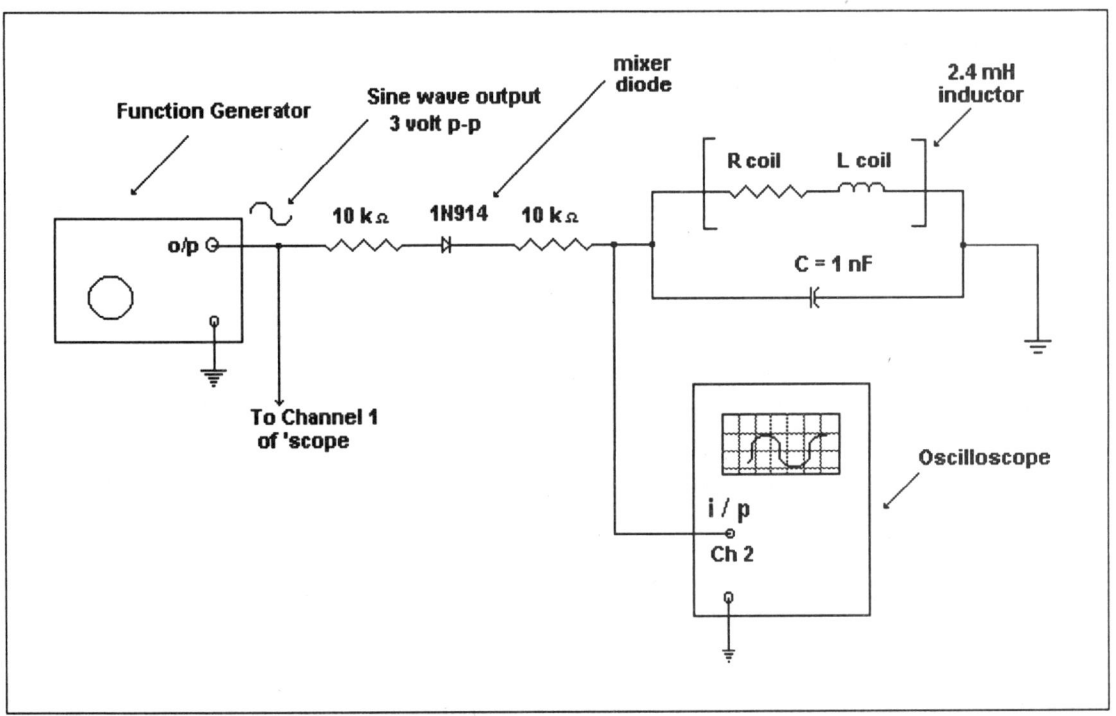

FIGURE 6-2

2. From the circuit values given, calculate the resonant frequency of the filter.

Answer: You should have calculated a resonant frequency of about 102.7 kHz.

Using channel 2 of the oscilloscope, monitor the voltage across the resonant circuit and carefully tune the frequency of the generator around the original setting of 50 kHz. As you tune you should observe the parallel resonant circuit go into resonance with the sudden appearance of a clean sine wave on the scope.

3. What is the frequency of the sine wave at the filter output?

4. Does the circuit provide frequency doubling?

Frequency Conversion and Mixer Action

Introduction

When two sine waves of differing frequencies, f_A and f_B, mix together in a non-linear circuit, the result is a complex mixture of sine waves of many different frequencies due to harmonic distortion and intermodulation distortion.

One of the effects of intermodulation distortion is the production of a sine wave whose frequency is equal to the difference between the frequencies of the two original sine waves. This is called a down conversion product. Another effect is the production of a sine wave whose frequency is the sum of the two frequencies. This is called an up conversion product.

The ability to move a signal to a different place in the spectrum is of central importance to communications systems.

An application of down conversion is found in a superheterodyne receiver where the incoming signal is down converted to a fixed frequency called the intermediate frequency by the use of a mixer circuit.

Procedure

Adjust the output of function generator A for 3 V p-p at 1.6 MHz. Adjust the output of function generator B for 3 V p-p at 1.5 MHz. Figure 6-3 is a conceptual diagram of the circuit setup.

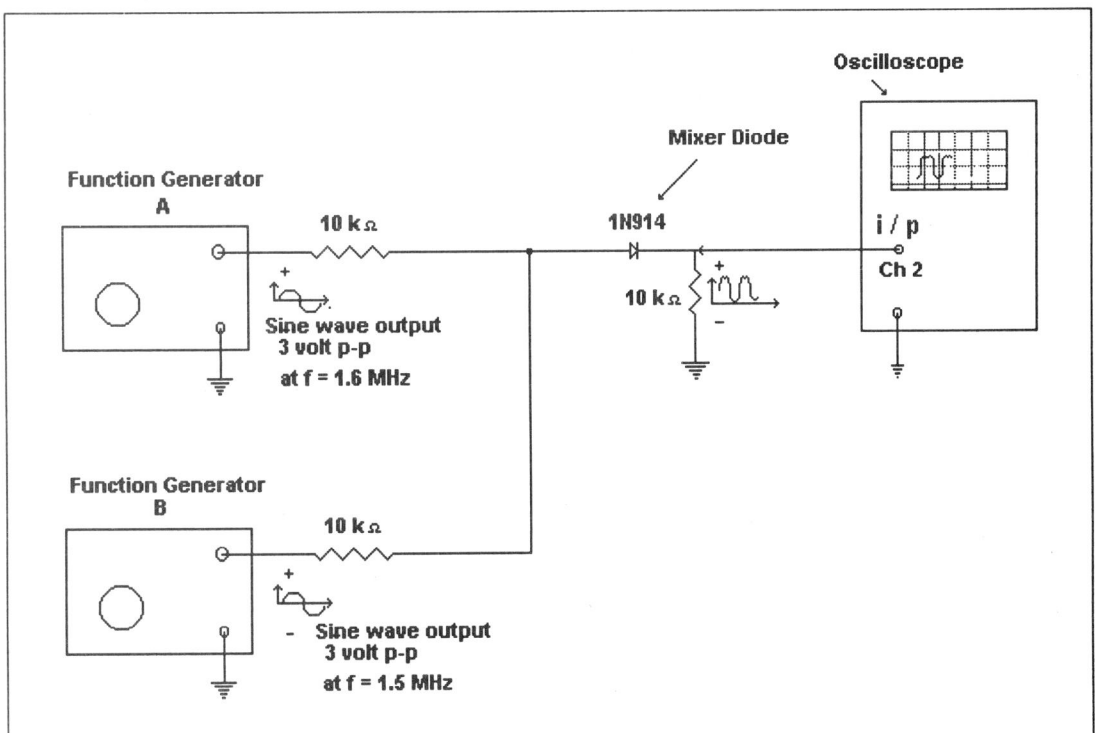

FIGURE 6-3

Use channel 1 and observe the resultant waveform at the input to the diode mixer. Adjust the oscilloscope controls for a stable display.

The signal at the input to the mixer is the combining of two equal amplitude sine waves separated in frequency by 100 kHz. The resulting waveform will be the addition and subtraction of the two phasors at a rate of 100 kHz. A cusping waveform should be observed.

5. Sketch the observed waveform.

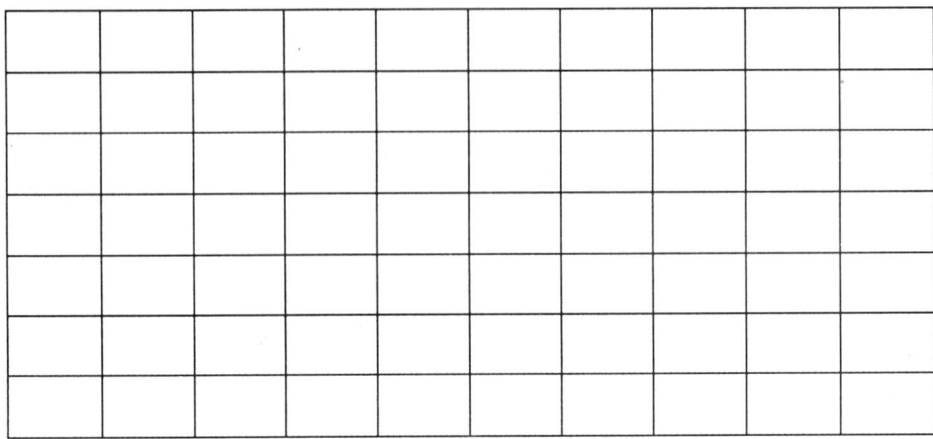

Use channel 2 and observe the waveform across the 10 kΩ load resistor.

The signal coming out of the mixer is distorted and contains a rich mixture of sine waves of many differing frequencies.

6. Roughly sketch the observed waveform.

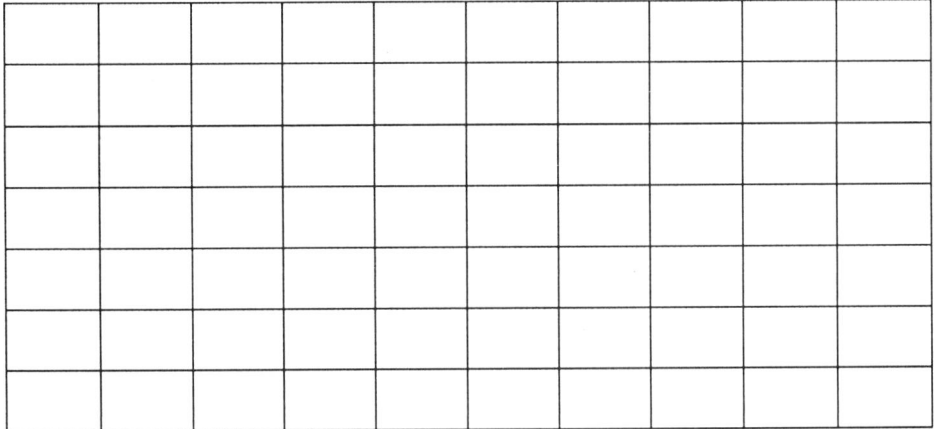

If the distorted signal coming from the mixer is filtered, it should be possible to isolate and observe the down conversion product.

where

$f_A = 1.6$ MHz

and

$f_B = 1.5$ MHz

Then the down conversion product will be at the frequency

$f_A - f_B = 100$ kHz

A parallel resonant circuit tuned to resonate close to 100 kHz will be used as a filter as shown conceptually in Figure 6-4. Connect the components as shown in Figure 6-4.

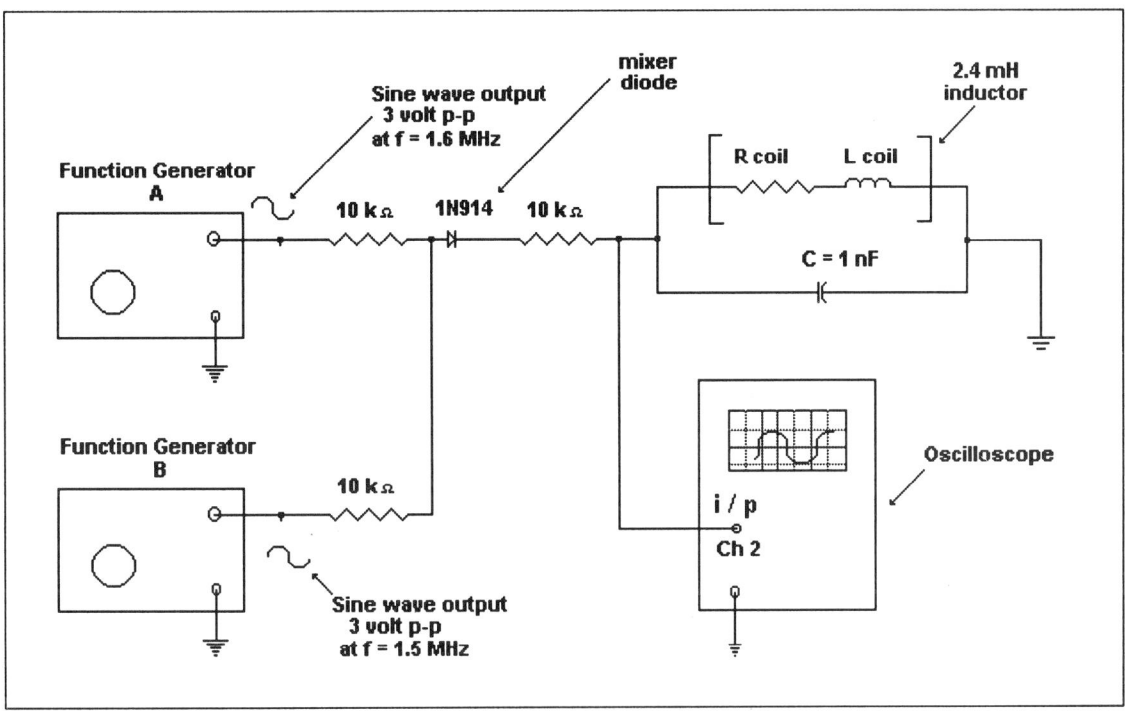

FIGURE 6-4

Carefully tune the frequency of generator A around the setting of 1.6 MHz and use channel 2 of the oscilloscope to monitor the voltage across the parallel resonant circuit. As you vary the frequency of generator A, you should observe strong resonance develop when the down conversion product is at the resonant frequency of the filter. A sine wave should appear on channel 2 of the scope. This should be the down conversion product from the mixer at a frequency close to 100 kHz.

Measure the exact frequency of this signal.

7. What is the measured frequency of the down conversion product?

NOISE

LAB 7

Name: _____   Date: _____

OBJECTIVES:

Upon completion of this lab, you will be able to:

- Measure the effect of channel bandwidth on noise power level
- Identify and use the video filter control on a spectrum analyzer
- Measure noise power levels with the aid of the video filter
- Measure the sensitivity of a spectrum analyzer
- Estimate the noise figure of a spectrum analyzer

TEST EQUIPMENT:

- Spectrum analyzer
- Optional: Connecting the video output from the spectrum analyzer to a large screen video monitor allows the demonstration of this material to a class. This would be a useful option if the spectrum analyzer is a scarce resource.

Procedure

> **NOTE:** A spectrum analyzer is easily damaged if the input signal is too large. Do not connect any signal source to the analyzer. If you are unsure about what you are doing, ask the instructor. **At no point in the lab should there be any signal source connected to the input of the analyzer.** At this time power should be off. Check that this is so.

Have the instructor check your work station before proceeding.

Instructor's signature: _____

Electronic Noise

Turn on the power to the spectrum analyzer. A screen should display information on the spectrum analyzer's settings. (On older analyzers this information has to be obtained from the knob settings.)

1. From the screen, find the following information.

    Reference level in dBm   = _____

    Vertical scale factor in dB/   = _____

    Center frequency in MHz   = _____

    Span in MHz/   = _____

    Resolution bandwidth in MHz   = _____

Notice that a fringe of noise (sometimes called "grass") is visible at the bottom of the screen.

The "grass" at the bottom of the screen is the visible display of noise at the output of the spectrum analyzer and is due to the various noise contributions made by all the various components (resistors, diodes, transistors, etc.) that make up the spectrum analyzer.

For example, a resistor produces thermal or Johnson noise, and a diode or transistor junction generates shot noise. When you consider the complexity and number of resistors and transistors in a spectrum analyzer, it is not surprising that the analyzer output display screen should have a noise signal on it.

2. Make a sketch of what you see. Indicate the reference level, scale factor, resolution bandwidth, and the approximate noise level in dBm on your sketch.

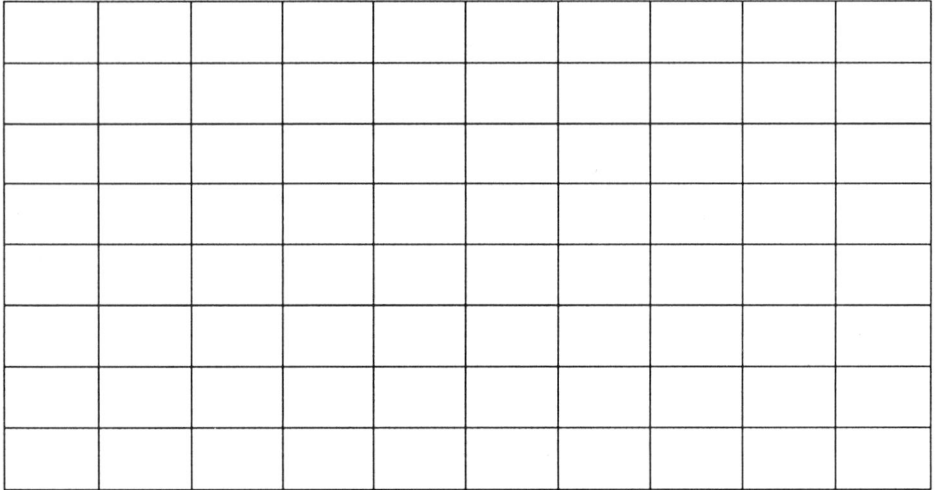

Noise and Sensitivity

The power level of the noise signal is of practical importance in determining how small a signal we can observe on the spectrum analyzer.

3. Make a *rough* estimate of the dBm level of the noise or "grass" at the bottom of the screen. (A noise signal is by nature amplitude erratic, so it is hard to get a definite reading.)

When a signal is connected to the spectrum analyzer input, the signal would not be seen on the analyzer display if the noise signal were the stronger of the two. Since a spectrum analyzer has the same architecture as a radio receiver, this statement is true for ultimately determining the ability of most receivers to detect weak signals.

4. Considering the dBm level of noise measured in step 3, what is the weakest signal that could be observed on the analyzer?

One of the basic concepts of communications theory is that if the bandwidth of a communications channel is made wider, the amount of information that can be transmitted increases. Unfortunately, so does the noise level. The reverse also happens. If the bandwidth is reduced, the noise level in a communications channel will go down. If it is important to receive a very weak signal, we can reduce the bandwidth and thereby reduce the noise level in a channel. The weak signal can be detected.

Try adjusting the span of the analyzer until the resolution bandwidth decreases by a factor of 10.

5. For a bandwidth change of 10, what is the change in the noise level in dB? (Again, make a rough estimate.)

 Answer: **You should have observed about a 10 dB change in the noise level.**

6. With reference to your text, explain how the theory of Johnson noise or thermal noise explains what you observed.

Reducing the resolution bandwidth of the spectrum analyzer should allow us to view a weaker signal than with the previous setting.

7. What is the smallest signal (roughly) that can now be observed?

Technologists who have to measure small signals that are nearly buried in noise use a secondary control of a spectrum analyzer called the video filter. In the previous measurements, it was difficult to measure the noise level on the analyzer since it was jumping about erratically.

In order to see the noise level more clearly, we will use the video filter control to measure the noise level more accurately. However, first we should develop a better idea of how a video filter works.

8. Refer to your text and write a short note explaining what a video filter is and its use.

Have your instructor show you the video filter control. Try using the video filter at several cutoff frequency settings and observe the result.

9. Make a rough sketch of what happens to the display of the noise level with and without the video filter.

As mentioned earlier, the smallest signal in dBm we can observe on any spectrum analyzer will be determined by the lowest level in dBm to which the noise level can be reduced. This is often descriptively called the "noise floor." This would also set the sensitivity specification for an analyzer or the minimum detectable signal (MDS).

The minimum detectable signal by an analyzer is dependent or referenced to such things as the settings of the input attenuator, resolution bandwidth, and video filter settings of the specific analyzer.

If it is available, the instructor will provide a copy of the page from the user manual for your analyzer detailing the test conditions for your specific analyzer.

10. From the specifications pages provided, locate as much as is stated on the test conditions for measuring the analyzer's sensitivity.

    Resolution bandwidth   = _____

    Video filter setting   = _____

    Input attenuator in dB   = _____

You should find that the resolution bandwidth specified will be as small as the analyzer will allow, and similarly for the attenuation setting. The video filter setting will typically be 100 times more narrow than the resolution bandwidth setting.

11. Adjust the spectrum analyzer for these settings (make sure there is no signal connected!) and make a measurement of the average noise level in dBm. This will give an estimate of the smallest signal (MDS) that can be detected by the analyzer.

Some corrections need to be applied to this measurement. The type of detector used in spectrum analyzers does not read noise signals accurately, and the logarithmic amplifier used to give a dB scale will also introduce inaccuracies. To compensate for this, a value of 2.5 dB is typically added to the average noise reading obtained with the video filter.

For example, if the average noise level were measured to be –95 dBm, then the corrected value would be found by adding 2.5 dB to the reading.

Corrected noise level = (–95 dBm) + (2.5 dB) = –92.5 dBm

12. What is the corrected MDS for the analyzer?

Noise Figure

The noise power at the display or output of the analyzer can be reduced by reducing the bandwidth of the analyzer, as observed earlier in the lab, or by reducing the noise figure of the spectrum analyzer. When the input attenuator was lowered, so was the noise figure of the analyzer.

If we make simplifying assumptions, the noise figure of the analyzer can be *approximately* estimated in the following way.

Suppose a 50 ohm signal source were to be connected to the 50 ohm signal input connector of the analyzer. **(Don't connect anything to the analyzer—just suppose you did.)** In such a case, the thermal noise power or Johnson noise power available from the 50 ohm signal source would flow into the analyzer.

The noise power from the 50 ohm signal source would not be seen on the analyzer screen, however, because the noise from the various components of which the analyzer is made (resistors, transistors, etc.) is much greater.

The signal power from the 50 ohm signal source at the analyzer input *would* be seen on the display of the analyzer if it were strong enough to overcome the effects of all the extra noise that the spectrum analyzer will contribute to the display. What is key in this kind of situation is the ratio of signal power to noise power.

The relationship between the signal-to-noise ratio at the spectrum analyzer input and the signal-to-noise ratio at the spectrum analyzer's output or display is

$$(S_o/N_o)_{dB} = (S_i/N_i)_{dB} - (NF)_{dB} \qquad (1)$$

If we were to turn down the imaginary signal generator power level until the signal-to-noise ratio at the display is 0 dB, then we could solve Equation 1 for the smallest signal detectable in terms of the analyzer noise figure.

$$0 = (S_i/N_i)_{dB} - (NF)_{dB}$$

or

$$(S_i)_{dBm} - (N_i)_{dBm} - (NF)_{dB} = 0 \qquad (2)$$

Therefore, the sensitivity or minimum detectable signal (*MDS*) of the analyzer in terms of the noise figure of the analyzer is

$$(MDS)_{dBm} = (N_i)_{dBm} + (NF)_{dB} \qquad (3)$$

The amount of power that would flow into the analyzer from the 50 ohm signal source is known from noise theory to be given by the following equation:

$$P_{\text{noise in dBm}} = -174 + 10 \log (B) \qquad (4)$$

where

B = noise bandwidth of the analyzer (Hz) ≈ 1.2 × analyzer resolution bandwidth

> *NOTE:* The definition of noise power bandwidth (*B*) used here assumes an ideal rectangular filter with the same power response as the actual instrument IF filter. Multiplying the resolution bandwidth by a correction factor of 1.2 gives an adequate estimate of the noise power bandwidth *B*.

Consequently, the relationship between sensitivity, noise figure, and bandwidth can be stated as

$$(MDS)_{dBm} = -174 + 10 \log (B) + (NF)_{dB} \qquad (5)$$

or

$$(\text{Noise Figure})_{dB} = +174 - 10 \log (B) + (MDS)_{dBm}$$

EXAMPLE CALCULATION: If the sensitivity of the analyzer measured in step 12 were −120 dBm and the resolution bandwidth of the analyzer were set at 1000 Hz, then the analyzer noise figure (*NF*) could be roughly estimated as

Analyzer NF in dB = +174 − 10 log (1.2 × 1000) + (−120) = 23.2 dB

---

13. Calculate the noise figure of your analyzer.

14. What is the significance of a 0 dB noise figure? Is it desirable?

15. Which of the following components is key in determining the noise figure of a superheterodyne receiver:

    RF amplifier

    Mixer

    IF amplifier

    Give reasons for your choice.

AMPLITUDE MODULATION WAVEFORM MEASUREMENTS

LAB 8

Name: _____   Date: _____

OBJECTIVES:

Upon completion of this lab, you will be able to:

- Measure the modulation index for an amplitude modulated (AM) signal
- Use the X vs. Y mode of the oscilloscope to observe the trapezoidal pattern for an AM signal
- Measure and interpret the spectrum of an AM signal

TEST EQUIPMENT:

- Dual trace oscilloscope with X vs. Y mode display
- RF generator with % and amplitude modulation input control
- Function generator
- Spectrum analyzer and (if available) large screen video display monitor

Procedure

In this lab, the AM signal to be tested will be generated at the instructor's bench by externally modulating an RF generator with the output of a function generator and distributed in parallel to the student work stations. Control of the modulated signal will rotate through the lab groups under the instructor's direction. The lab setup is shown conceptually in Figure 8-1.

The following are the test conditions for this part of the lab:

- Function generator is set to 400 Hz. This sine wave signal will be the intelligence or modulating signal and will be connected to the external modulation input control of the RF generator as shown in Figure 8-1. The output level needed from the function generator will depend on the specific characteristics of the RF generator external modulation input control.

- The RF generator is set to output a 60 kHz signal with a 2 V p-p output. This signal will be the carrier frequency.

- The RF generator modulation index control is set to produce an amplitude modulated waveform with a modulation index of 30%.

AM Waveform

Connect the 400 Hz modulating signal to channel 1 of your oscilloscope. Connect the amplitude modulated signal to channel 2 of your oscilloscope. For a stable display, trigger the oscilloscope on channel 1.

With the time base set to 1 ms/div, you should see an AM waveform with a modulation index of 30% (m = 0.3).

40 LAB 8 AMPLITUDE MODULATION WAVEFORM MEASUREMENTS

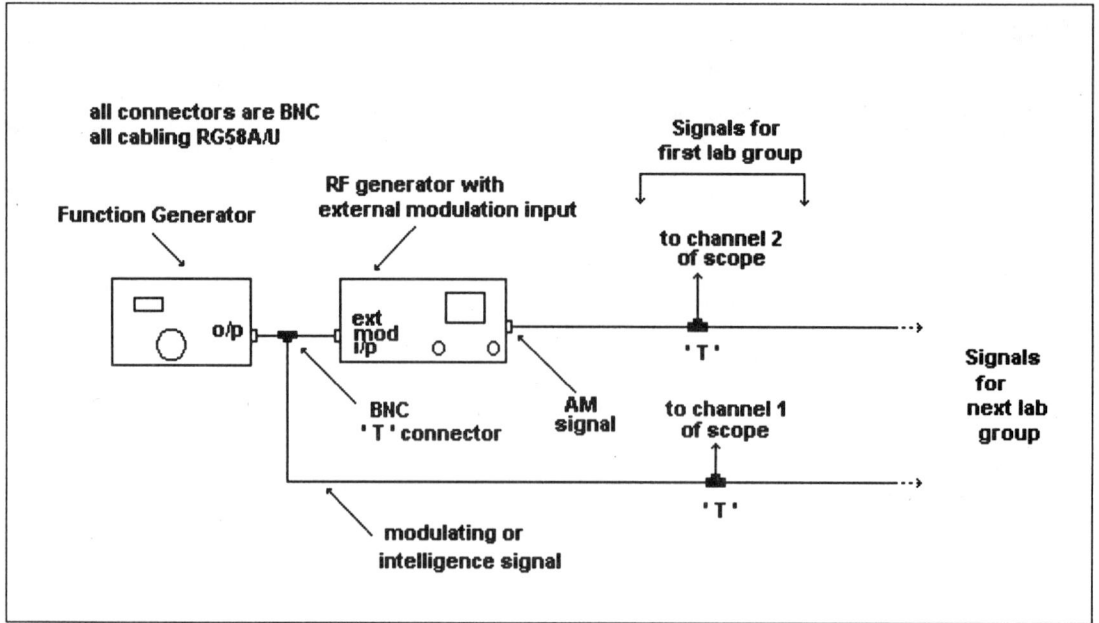

FIGURE 8-1

1. Sketch the waveforms observed on the oscilloscope. Show voltage levels and observed times for your sketches.

The modulation index for an AM waveform can be calculated from the following equation, using the pk-pk voltage of the waveform peaks as E_{max} and the pk-pk voltage of the waveform troughs as E_{min}.

$$\%m = \frac{E_{max} - E_{min}}{E_{max} + E_{min}} \times 100\%$$

2. Show by calculation that $m = 30\%$ for the observed AM signal.

Trapezoidal Pattern

An important test pattern used in monitoring an amplitude modulation signal is the trapezoidal pattern. To view the trapezoidal pattern for this AM signal:

- put the scope into X-Y mode
- connect the intelligence signal to the horizontal input of the oscilloscope
- connect the AM signal to the vertical input of the oscilloscope
- adjust the oscilloscope controls for a nicely centered display

Figure 8-2 shows the display of a trapezoidal pattern for an AM signal.

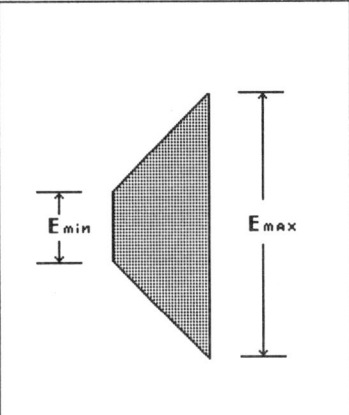

FIGURE 8-2

3. Sketch your display. Using the values of E_{max} and E_{min} obtained from your oscilloscope display, calculate the modulation index.

Leave the X-Y mode and reset the oscilloscope settings to redisplay the AM waveform.

Adjust the modulation index for a mod index of 0%. Measure the E_{max} of the unmodulated carrier. Since there will be no peaks and troughs, this will just be the peak-to-peak voltage of the 50 kHz carrier which was set at the start of the lab to be 2 volts pk-pk.

Adjust the modulation index for a mod index of 50%. Measure the E_{max} of the modulated carrier.

4. What was the amplitude of the unmodulated carrier?

5. What was the E_{max} of the signal with a modulation index of 50%?

6. What is the E_{min} of the signal with a modulation index of 50%?

You should have noticed that a modulation index of 50% caused the amplitude of the modulated signal to swing up 50% to 3 volts pk-pk and down 50% to 1 V pk-pk.

Adjust the modulation index to 80% and observe the modulated waveform.

7. What values would you expect for E_{max} and E_{min} in volts pk-pk?

8. What did you measure for E_{max} and E_{min}?

Signal with Unknown Modulation Index and Unknown Modulating Frequency

In this part of the lab the modulation index and modulating frequency are set by the instructor to values unknown to the lab groups. The lab groups will use both the normal oscilloscope waveform display of the AM signal and the X-Y mode of the oscilloscope to measure the % modulation index and modulating frequency.

Use the same method as before to observe the AM waveform on the oscilloscope.

9. Indicating voltage levels for E_{max} and E_{min} and times, sketch the observed AM waveform on the display provided. From the measured values of E_{max} and E_{min}, calculate the modulation index and the frequency of the modulating signal.

Use the same method as before to obtain the oscilloscope X-Y display of the AM waveform.

10. Sketch the resulting trapezoid display. From measurements of the trapezoid, calculate the % modulation index.

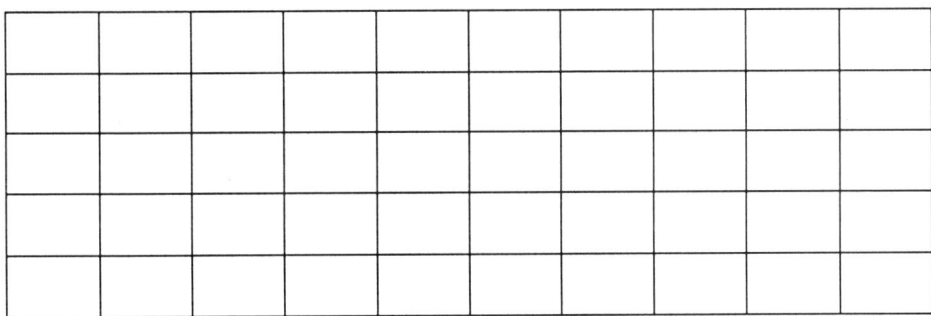

AM Spectrum

This part of the lab will be performed by the instructor as a demonstration.

The output of the RF generator is set for an output of 0 dBm at 10 MHz. It is then externally modulated with a signal from the function generator of 10 kHz with the modulation index adjusted for 50%. The AM signal is connected to the spectrum analyzer input and (optionally) the video signal output of the spectrum analyzer is connected to the input of a large screen video display monitor. The resulting displayed spectrum is to be used in a group discussion and to gather data for further calculations.

Observe the spectrum of the AM signal on the analyzer screen.

11. Sketch the resulting display showing the analyzer settings for center frequency, reference level, etc.

The spectrum displayed is the result of modulating a 10 MHz carrier signal with a 10 kHz sine wave. When a carrier is modulated with a single sine wave, the result is three RF signals.

From AM theory, you should remember that there will be the original carrier signal and two side frequencies—an upper side frequency and a lower side frequency.

12. What, in theory, should be the frequencies of these signals?

13. What do you measure on the screen of the spectrum analyzer?

You should have seen the original carrier signal at 10 MHz, two relatively strong side frequencies at 10.01 MHz and 9.99 MHz, and possibly numerous weaker side frequencies due to the 10 kHz modulating sine wave not being a pure sine wave. As a consequence, the harmonics of the 10 kHz sine wave also modulate the carrier, producing extra sets of side frequencies around the carrier. In the following calculation we will ignore these side frequencies.

The power in each of the two relatively strong RF side frequencies is mathematically related to the power in the carrier. Remember the power in the carrier is set to 0 dBm (1 milliwatt) and the modulation index is set to 50%.

14. What power in dBm should there be in the carrier signal displayed on the screen? What did you measure?

15. What power in dBm should there be in each side frequency? Show your calculations.

16. How well do your calculations compare with the measured values?

The modulating waveform will now be changed to a 10 kHz square wave with no changes to any of the other settings.

The analyzer display should be affected by the harmonic content of the square wave.

17. Sketch the resulting AM spectrum. Using your knowledge regarding the spectral content of a square wave, explain what you observe.

RECEIVER SENSITIVITY

LAB 9

Name: _____ Date: _____

OBJECTIVES:

Upon completion of this lab, you will be able to:

- Measure the sensitivity of an AM receiver and compare the measurements to the manufacturer's specifications
- Measure the gain of a receiver
- Measure the total harmonic distortion of a receiver and compare the result to the manufacturer's specifications
- Monitor the AGC of the receiver, measure the AGC figure of merit, and compare the result to the manufacturer's specifications

TEST EQUIPMENT:

- Citizen's Band (CB) radio with the following specifications and to be operated in this lab with the following operating constraints:
  1. The CB radio is designed and manufactured to be operated *only* from a low voltage of +13.8 volts dc (nominal) and will be powered in this lab by a low dc voltage of +13.8 volts dc from an adjustable low voltage lab bench dc power supply.
  2. The CB has a 40-channel capability, each channel 10 kHz wide, in the range from 26.965 MHz to 27.405 MHz with a maximum transmitted power in AM mode of 4 watts and 12 watts peak envelope power (PEP) in SSB.
  3. The CB unit is to be operated only in receive mode throughout the lab (a microphone will not be issued) and has a maximum audio output power of 3 watts into an 8 Ω load.
  4. A microphone is not to be part of the lab equipment, so choose a CB that can operate in receive mode when the microphone is not connected to the radio (UNIDEN model PC122XL or equivalent).
- External 8 Ω speaker box—President Model 711-SX or equivalent and audio cabling
- RF generator
- True RMS ac voltmeter
- Distortion analyzer
- Adjustable low voltage dc power supply (GW model GPC-3020 or equivalent)

Procedure

Figure 9-1 illustrates conceptually the equipment setup. Your instructor will provide you with a detailed diagram for connecting the specific equipment at your lab station. Using the instructor-provided diagram, connect the equipment but **do not turn on any power** until the instructor has checked your equipment setup. Microphones will not be issued or connected to avoid accidentally transmitting into the signal generator.

NOTE: It is dangerous to connect an antenna to a receiver near power lines or in confined spaces such as in a lab. At no time in this lab should there be an antenna hooked to the transceiver.

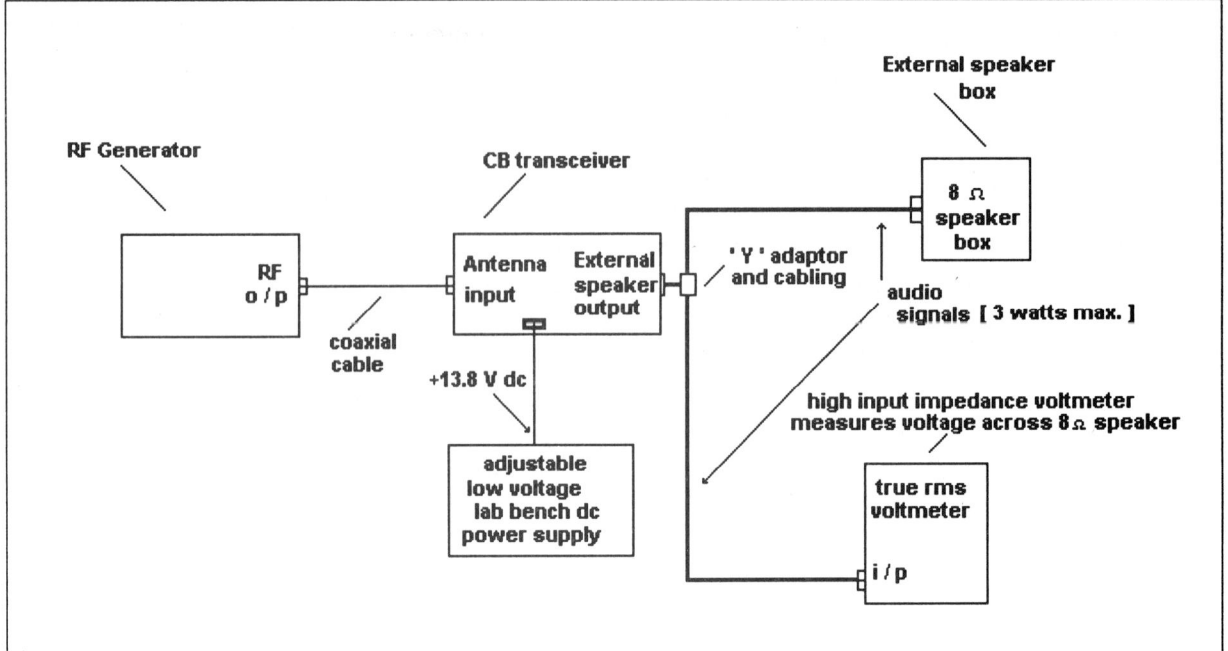

FIGURE 9-1

 Instructor's signature: _____

Now turn on the power for your CB (nominally +13.8 V dc from an adjustable low voltage bench dc power supply). Using the instruction manual for your CB transceiver, set the transceiver controls as follows:

- AM
- Squelch fully C.C.W.
- Any convenient channel—we will use ch 25 (27.245 MHz)
- Full sensitivity (all RF stages ON)

Receiver Sensitivity Specifications

Any receiver has a specified sensitivity which tells the user what is the smallest usable signal that can be received. For example, the sensitivity of a receiver might be quoted on the spec sheets as 0.7 μvolts for a 10 dB (signal + noise)-to-noise ratio. This tells us that if an RF signal of 0.7 μvolts is received on the input of the receiver, the audio output (signal + noise)-to-noise ratio from the receiver will be at least 10 dB.

Attached to this lab or supplied with the CB instruction manual you are working with is a specification sheet for your transceiver.

1. From this spec sheet, what is the sensitivity rating?

What is often left unstated in these sensitivity specifications is the absolute audio output power at which the (signal + noise)-to-noise ratio is measured. There is no point in having a very sensitive receiver that is not capable of being heard. For our purposes in this lab, we will assume that we will measure the rated sensitivity of our receiver when the audio stage is operating at half the maximum rated power.

2. What do the spec sheets say is the maximum audio output power for your receiver?

A typical value would be 3 watts. The value you located on the spec sheets will probably be accompanied with a distortion specification of perhaps 10%. What the manufacturer is saying is that the receiver can output 3 watts and still have an output with a distortion of 10% or less.

3. Can you locate a distortion specification to accompany the maximum audio rated output power? If so, what is the value?

In the following measurements of receiver sensitivity, we will measure the sensitivity while maintaining the output power of the audio stage at half of the rated output power.

Measure the AM sensitivity as follows:

(a) Set RF generator for the sensitivity value found in question 1 (typically about 0.7 μvolts) with 30% modulation @ 1000 Hz and at the same frequency to which your receiver is tuned. If you set your transceiver to channel 25 (at 27.245 MHz), then set the RF generator to this frequency. Use the frequency control vernier to fine tune the RF output frequency. You will know you are right on ch 25 when the reading on the RMS voltmeter peaks.

(b) Adjust the audio frequency (AF) gain (volume) control until the audio output power is at half the rated power. If the rated power in question 2 were 3 watts, then half power would be 1.5 watts. Since the speaker has an 8 Ω impedance, the voltage observed on the true rms voltmeter will be as follows:

$$V = \sqrt{PR}$$
$$= \sqrt{(1.5\,W)(8\,\Omega)} = 3.46\,V$$

(c) Now turn off the modulation on the RF generator. The change in the reading on the rms voltmeter should be 10 dB or more. If the change is more than 10 dB, decrease the RF generator output voltage. That is, change the RF generator signal from 0.7 μV to 0.6 μV and remeasure the change in the audio output as modulation is applied and removed. Repeat this step until the output of the RF generator gives a 10 dB change as modulation is removed. (Note: If the rms voltmeter does not have a dB scale, use 20 log (audio voltage change) to measure the 10 dB point.)

4. What was the measured sensitivity in μV?

Receiver Gain

Determine the receiver gain as follows:

$$A_P \text{ (dB)} = A_V \text{ (dB)} + 10 \log \frac{Z_i}{Z_o}$$

$$\text{Receiver gain (dB)} = 20 \log \frac{3.46 \text{ V}}{\text{sensitivity}} + 10 \log \frac{50 \, \Omega}{8 \, \Omega}$$

5. What is the receiver gain?

It is important to realize that the gain of a receiver is a variable and that when the signal on the input of the receiver is tiny, the receiver automatic gain control will boost the gain of the receiver. A value of about 140 dB will not be untypical. If the input signal were to increase, the gain of the receiver would decrease radically.

Distortion

Previously in this lab we found the distortion specification for the receiver (question 3).

A distortion specification for a typical receiver might read as being 10% Total Harmonic Distortion (THD) @ 1000 µV input signal with 30% modulation @ 1000 Hz when putting out 3 watts (max) output.

We will measure the distortion when the receiver is operating at maximum rated output audio power and when the RF input signal is 1000 µvolts modulated with 1000 Hz @ 30% modulation index.

(a) With the equipment connected as shown in Figure 9-1, connect the distortion analyzer and the rms voltmeter to the audio output using a T-connector. This will allow both the rms voltmeter and the distortion analyzer to be displayed at the same time.

(b) Set the RF generator for 1000 µV @ 30% modulation @ 1000 Hz at the receive frequency of your transceiver. Fine tune the O/P frequency of the RF generator again by tuning the RF frequency vernier control and peaking the rms voltmeter readings. Using the AF (volume) gain control, adjust the audio output to the rated maximum audio output power. As an example, if the rated output maximum power was 3.0 watts, the rms voltmeter reading across the 8 Ω external load would be

$$V = \sqrt{PR}$$
$$= \sqrt{(3.0 \text{ W})(8 \, \Omega)} = 4.90 \text{ V}$$

(c) Bring the audio signal from the test-jig to the distortion analyzer input and measure the distortion in % THD. Instruction sheets for using the distortion analyzer are attached to this lab or will be provided by your instructor.

6. What is the measured distortion for the receiver?

7. Did the receiver meet spec?

AGC Characteristics

Automatic gain control (AGC) is used to keep the audio output of a receiver from varying too much as the input RF signal is changed. The AGC figure for a receiver is characterized in different ways by different manufacturers. One method of specifying the AGC capability of a receiver is the *AGC figure of merit*. Suppose the transceiver you are testing specifies an AGC figure of merit of 80 dB. This figure is derived from the following relationship:

$$\text{AGC figure (dB)} = 20 \log \left[\frac{\text{RF INPUT maximum/minimum}}{\text{AUDIO OUTPUT maximum/minimum}} \right]$$

EXAMPLE: If the RF input to a receiver is varied from 100,000 µV to 1 µV with the audio output level varying from 6V to 3V, this would give a figure of merit of

$$\text{AGC figure (dB)} = 20 \log \left[\frac{\frac{100 \text{ mV}}{1 \text{ µV}}}{\frac{6 \text{ V}}{3 \text{ V}}} \right]$$

$$= 20 \log \left(\frac{100 \text{ mV}}{1 \text{ µV}} \right) - 20 \log \left(\frac{6 \text{ V}}{3 \text{ V}} \right)$$

$$= 100 \text{ dB} - 6 \text{ dB} = 94 \text{ dB}$$

Expressed in another way, the AGC figure is found by subtracting the audio level change in dB from the RF input level change in dB.

Test the transceiver as follows:

(a) Set up the equipment as shown in step 1 when measuring sensitivity.

(b) Set the RF signal generator output to 100 mV.

(c) Modulate the signal to a level of 30% at 1 kHz.

(d) Adjust the audio output power of the receiver for half power.

(e) The signal generator level is incrementally adjusted from 100 mV to 1 µV. Record the data in Table 9-1.

Table 9-1

RF input (V)	Audio output (V)
100 mV	
30 mV	
10 mV	
3 mV	
1 mV	
300 µV	
100 µV	
30 µV	
10 µV	
3 µV	
1 µV	

Plot the audio output versus input signal level input on graph paper from 1 µV to 10 mV. The resulting curve should be smooth and free of abrupt changes.

8. What is the measured AGC figure of merit in dB?

9. What is the gain of the receiver when the input signal is 1 µV?

10. What is the gain of the receiver when the input signal is 100 mV?

Notice the radical change in gain as the input signal is varied.

BALANCED MODULATORS

LAB 10

Name: _____ Date: _____

OBJECTIVES:

Upon completion of this lab, you will be able to:

- Plot the output frequencies of a balanced modulator given the input RF signal and the input local oscillator signal
- Observe how a balanced modulator can be used to upconvert a signal frequency
- Observe and measure the carrier suppression at the output of the balanced modulator

TEST EQUIPMENT:

- Two function generators
- Oscilloscope
- Spectrum analyzer with the option of the video output connected to a large screen video monitor. The use of a video monitor allows the display of the small screen of the spectrum analyzer to a class in a demonstration format.
- Balanced modulator (Mini-Circuits SBL-1)

Pre-Lab

To perform this lab, the student should have received prior instruction on mixers and balanced modulators.

This lab centers around the testing of the SBL-1 balanced modulator, and the specification sheets are provided in Appendix A. You will find on these sheets a schematic of the SBL-1 balanced modulator (see specification sheet 1-37 in Appendix A).

As you can see from the diagram, there is an arrangement of four diodes in the package. This type of balanced modulator is called a double balanced diode mixer. It is also sometimes called a ring modulator.

1. Explain the operating principles of a double balanced diode mixer. Include a diagram and any math that is helpful to your explanation. (Use your text or suitable reference for your research.)

Balanced Modulator Specifications

To familiarize ourselves with typical balanced modulator specifications, imagine that the equipment in Figure 10-1A has been hooked up and we have the resulting display in Figure 10-1B on the spectrum analyzer. The numbers shown here are only to illustrate concepts and will not be the values you will actually use when we test the SBL-1 balanced modulator.

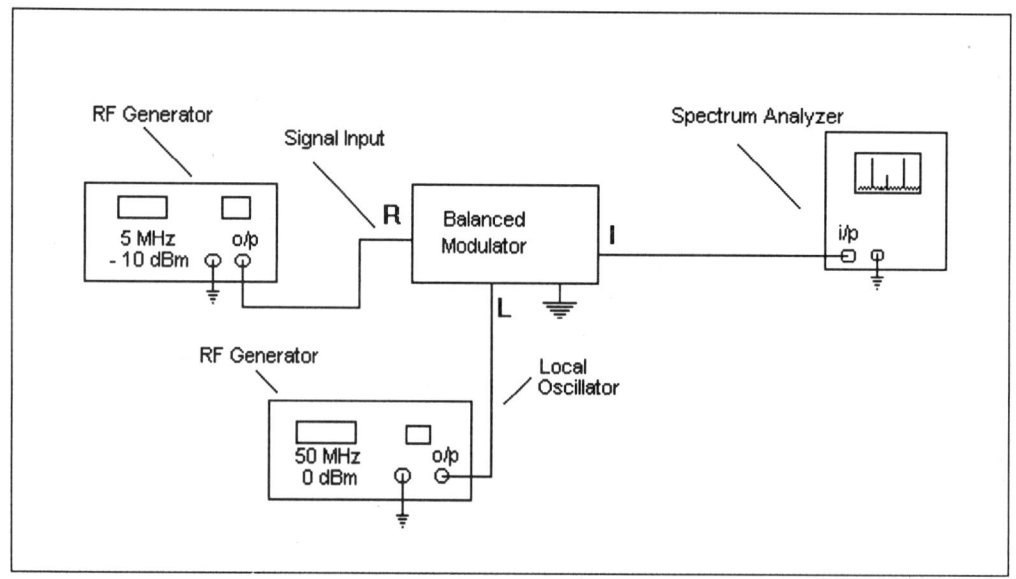

FIGURE 10-1A

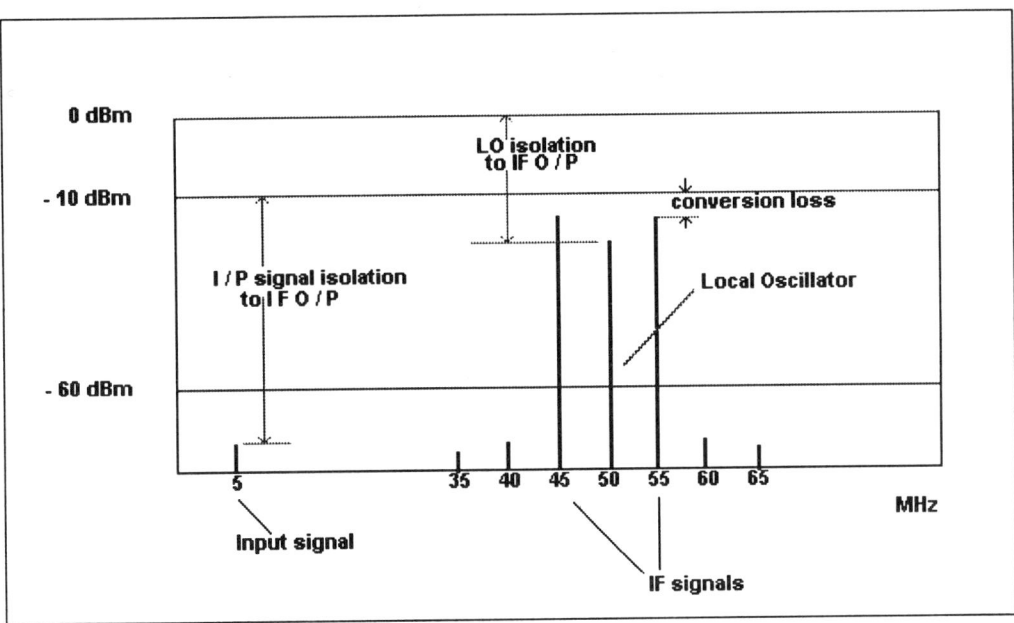

FIGURE 10-1B

As can be seen from the diagram, the signal to be upconverted comes into the balanced modulator input at port R at a frequency of 5 MHz and a strength of −10 dBm and is output from the balanced modulator upconverted to 45 MHz and 55 MHz. Note that these upconversion products have signal strengths that are less than −10 dBm when they come out of the I port of the balanced modulator. The frequency conversion process has caused a *conversion loss*.

Another important specification is the degree to which the local oscillator signal is suppressed by the balanced modulator. The local oscillator signal in Figure 10-1A has a strength of 0 dBm at port L of the balanced modulator but has a signal strength on the analyzer display which is much less than 0 dBm. Ideally we would like the 50 MHz signal to not even register on the analyzer display. The specification that deals with this is the *LO-IF isolation* and it is quoted in dB.

Reference the specifications for this balanced modulator and answer the following questions for the SBL-1 balanced modulator.

2. What is the maximum conversion loss in dB for the SBL-1?

3. What is a typical mid-band LO-IF isolation value in dB?

Procedure

The following procedure references the conceptual diagram in Figure 10-2.

Turn on function generator A before connecting it to the balanced modulator and use the oscilloscope to set its output to 0.5 volts p-p at 100 kHz.

Turn on function generator B before connecting it to the balanced modulator and use the oscilloscope to set its output to 1 V p-p at 2 MHz.

Connect the circuit shown conceptually in Figure 10-2.

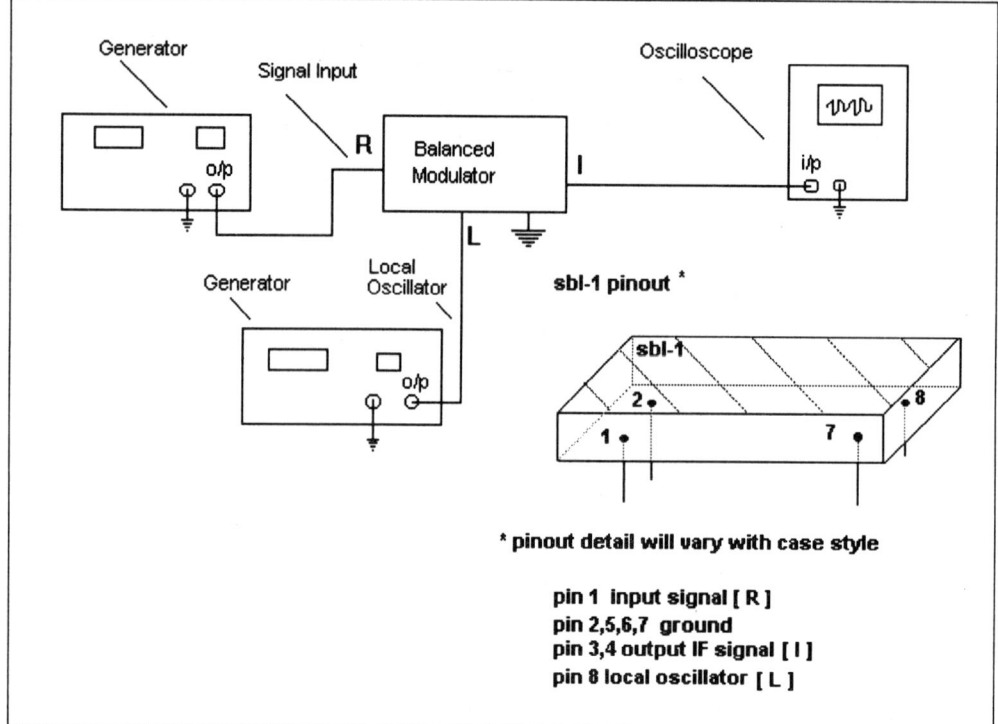

FIGURE 10-2

Adjust the oscilloscope controls and observe the waveform that is being produced.

Since the 2 MHz signal is in effect multiplying the 100 kHz alternately by +1 and then by −1, at a rate of 2 MHz you should observe a 100 kHz signal that is being chopped up at this rate—alternately going positive and then negative.

4. Sketch the observed signal.

Turn off the power to your equipment and replace the oscilloscope with the spectrum analyzer as shown conceptually in Figure 10-3. Do not turn on any power until the instructor has checked your setup.

 Instructor's signature: _____

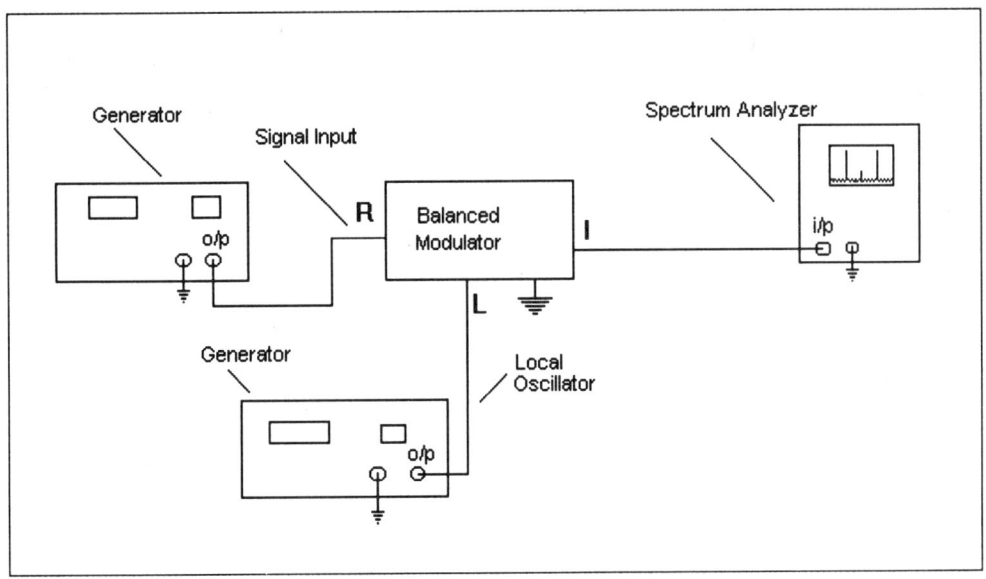

FIGURE 10-3

Now turn on the power for the equipment. Adjust the spectrum analyzer for a span of 500 kHz/div and a center frequency of 2 MHz. Adjust the reference level so the top graticle line is initially 0 dBm.

5. Plot the output of the balanced modulator as it is displayed on the spectrum analyzer. Indicate the key analyzer settings on the plot.

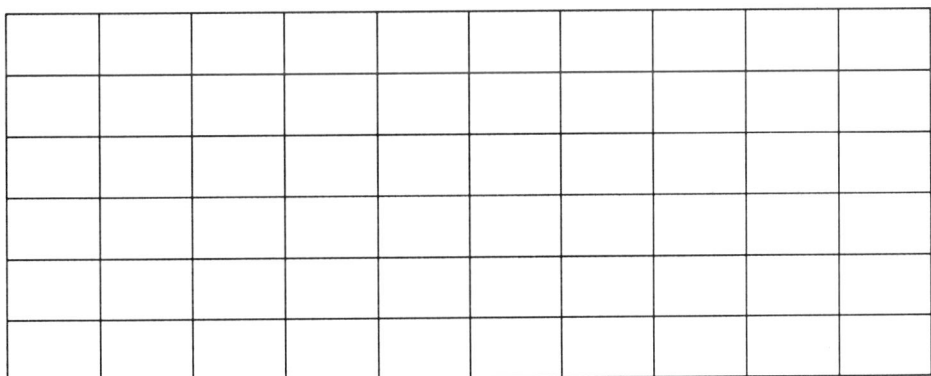

An examination of the analyzer display should show the suppressed carrier and two relatively strong (compared to the carrier) modulation products as well as weaker higher-order modulation products.

6. What is the strength in dBm of the two upconverted modulation products on either side of the carrier?

7. What are their frequencies? Are they where you predicted they would be?

8. What is the strength of the carrier in dBm? What is its frequency?

9. How much weaker in dB is the carrier compared to the upconverted modulation products?

SSB FUNDAMENTALS

LAB 11

Name: _____ Date: _____

OBJECTIVES:

Upon completion of this lab, you will be able to:

- Measure the sensitivity of an SSB receiver and compare the measurements to the manufacturer's specifications
- Compare and explain the difference in sensitivity for SSB operation and AM operation
- Manipulate and explain the operation of the clarifier
- Measure the squelch threshold and observe squelch action

TEST EQUIPMENT:

- Citizen's Band (CB) radio with the following specifications and to be operated in this lab with the following operating constraints:
  1. The CB is designed and manufactured to be operated *only* from a low voltage of +13.8 volts dc (nominal) and will be powered in this lab by a low dc voltage of +13.8 volts dc from an adjustable low voltage lab bench dc power supply.
  2. The CB has a 40-channel capability, each channel 10 kHz wide, in the range from 26.965 MHz to 27.405 MHz, with a maximum transmitted power in AM mode of 4 watts and 12 watts peak envelope power (PEP) in SSB.
  3. The CB radio is to be operated only in receive mode throughout the lab (a microphone will not be issued) and has a maximum audio output power of 3 watts into an 8 Ω load when powered by +13.8 volts dc.
  4. A microphone is not to be part of the lab equipment, so choose a CB that can operate in receive mode when the microphone is not connected to the radio (UNIDEN model PC122XL).
- External 8 Ω speaker box—President Model 711-SX or equivalent and audio cabling
- RF generator
- True RMS voltmeter
- Adjustable low voltage dc power supply (GW model GPC-3020 or equivalent)

Procedure

Figure 11-1 illustrates conceptually the equipment setup. Your instructor will provide you with a detailed diagram specific to the particular models of equipment available in your lab. Set up the equipment as detailed in the diagram provided by the instructor, but **do not turn on the power** to the equipment until the instructor has checked your setup. A microphone will not be part of the lab equipment to avoid accidentally transmitting into the signal generator.

➤ **NOTE:** It is dangerous to connect an antenna to a receiver near power lines or in confined spaces such as in a lab. At no time in this lab should there be an antenna hooked to the transceiver.

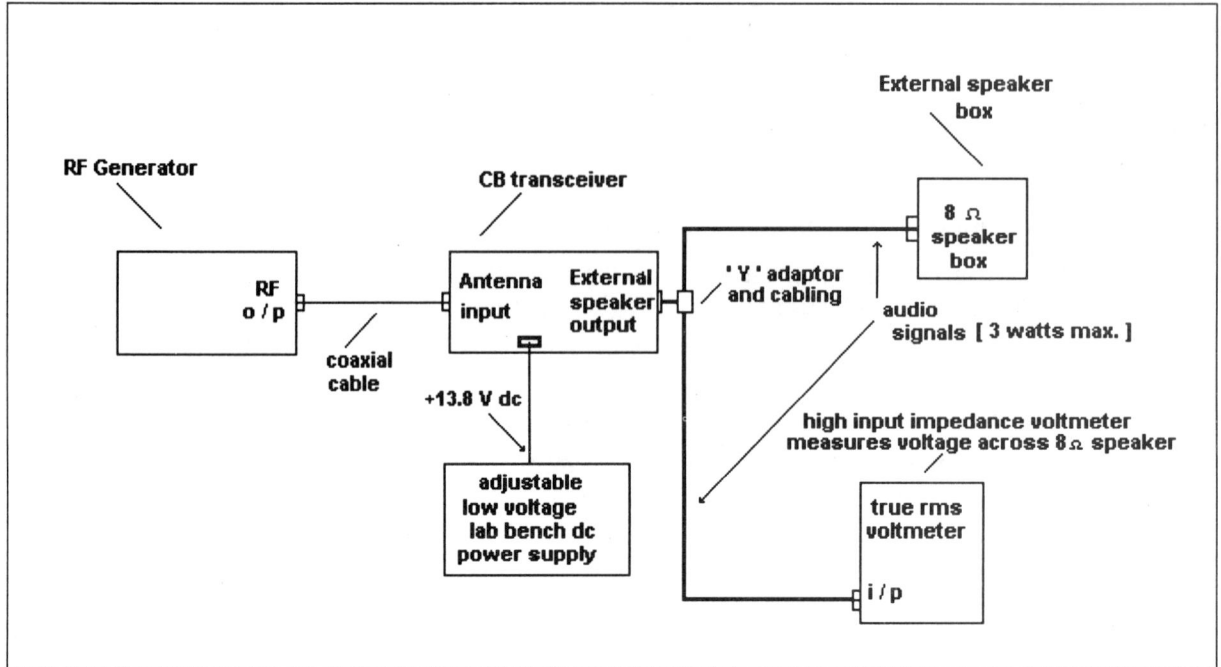

FIGURE 11-1

 Instructor's signature: _____

Now turn on the power supply for the CB (+13.8 V dc from an adjustable low voltage bench power supply).

SSB Sensitivity Measurement

The overall objective of this measurement is to find what is the smallest RF signal at the receiver input that will give a 10 dB (S+N)/N [(signal + noise)-to-noise] ratio at the speaker output while operating the receiver at half its rated power.

Consult the specification sheets for the transceiver (provided by the instructor at the beginning of the lab) and locate the 10 dB (S+N)/N sensitivity specification for your transceiver.

1. What is the specified 10 dB (S+N)/N sensitivity in μvolts?

To measure the 10 dB (S+N)/N sensitivity specification, perform the following steps:

    (a) Set the receiver for upper side band operation (USB) on a suitable channel such as channel 25 (27.245 MHz).

    (b) Set the squelch fully counterclockwise (ccw). If the receiver has a "full sensitivity" mode such as HWY mode, ensure this is on.

    (c) Set the RF generator for an output of 1.0 µV. Make sure the modulation is off.

    (d) Set the frequency of the RF generator to a frequency that will lie in the passband of your receiver. For a receiver set to receive in upper side band on channel 25, this would be at about 27.246 MHz.

    (e) Fine tune the frequency of the RF generator until you can hear a tone in the receiver. Adjust the frequency for a peak reading on the true-RMS voltmeter. At this point it does not matter what the audio output level is. We are just trying to set the RF generator to the correct frequency.

    (f) Adjust the audio frequency (AF) gain control until the audio output power is half the rated maximum audio output power. For example, for a half power output of 1.5 watts into 8 Ω, the voltmeter should be indicating 3.48 volts.

2. Show how this 3.48 volt reading is calculated.

    (g) Switch the receiver to LSB. This has the same effect as removing the signal but not the noise.

    (h) Note the dB change as indicated on the voltmeter dB scale. This change is the difference between the signal + noise power level and the noise power level.

    (i) If the measured change was 10 dB, then the RF generator signal level is set right on the sensitivity rating of the receiver. If the dB change is not 10 dB, adjust the RF generator output and repeat the above measurement steps until a 10 dB change in audio output power is obtained.

3. What is the 10 dB (S+N)/N value for SSB?

    _____ µV

4. Did the measured sensitivity meet the manufacturer's spec?

AM Sensitivity Measurement

Perform an AM sensitivity measurement and compare it with the SSB sensitivity measurement. The procedure is outlined below.

Connect the equipment as illustrated in Figure 11-1. To measure the 10 dB S+N/N sensitivity, proceed as follows:

(a) Set the receiver for AM operation on a suitable channel such as channel 25 (27.245 MHz).

(b) Set the squelch level fully ccw and if the receiver has a "full sensitivity" mode, ensure it is on.

(c) Set RF generator for 0.7 µV @ 30% modulation @ 1000 Hz and on ch 25 (27.245 MHz) of your transceiver. Make sure the receiver is set to full sensitivity mode. Use the frequency control vernier to fine tune the RF output frequency. You will know you are right on ch 25 when the reading on the RMS voltmeter peaks.

(d) Adjust the audio frequency (AF) gain control until the audio output power is 1.5W. Since the speaker has an 8 Ω impedance, the voltage observed on the true rms voltmeter will be as follows:

$$V = \sqrt{PR}$$
$$= \sqrt{(1.5\,W)(8\,\Omega)} = 3.46\,V$$

(e) Now turn off the modulation by switching from internal to external modulation on the RF generator. The change in the reading on the rms voltmeter should be 10 dB or more. If the change is more than 10 dB, decrease the RF generator output voltage. That is, change the RF generator signal from 0.7 µV to 0.6 µV and remeasure the change in the audio output as modulation is applied and removed. Repeat this step until the output of the RF generator gives a 10 dB change as modulation is removed.

5. What is the AM sensitivity measurement?

   _____ µV

6. How does the AM sensitivity of your set compare to the SSB sensitivity of your receiver? In which mode do you have the better sensitivity?

Answer: **You should have noticed a better sensitivity in the SSB mode.**

7. What explanation can you give for this result?

Clarifier Operation

In SSB operation it is imperative that the beat frequency oscillator is accurate. If the oscillator is out by so much as 100 Hz, the audio will sound "funny" and may even be unintelligible. To ensure proper sounding audio, the designers of SSB transceivers include a clarifier. Quite often the clarifier is a variable capacitor that can change the frequency of the beat frequency oscillator by a small amount.

8. Describe the clarifier circuit used in your CB transceiver. Refer to the block diagram for the transceiver found in the CB user's manual or provided by your instructor.

To observe the effect the clarifier has on the received signal, perform the following steps:

(a) Set the transceiver to USB operation.

(b) Set the RF generator for an output of 1 µV. Make sure the modulation is off.

(c) Adjust the RF generator frequency until a tone is heard in the speaker.

(d) Vary the clarifier and observe how the audio frequency changes.

Squelch Circuit

The squelch circuit prevents any audio from reaching the audio amplifiers until a signal of sufficient strength is received. Mobile radios would be intolerable if a radio operator had to listen to radio noise the whole time.

To become familiar with the operation of the squelch circuit, perform the following steps:

(a) Turn the squelch off, completely counterclockwise.

(b) Set the transceiver up for USB on channel 25.

(c) Adjust the RF signal generator for about a 1000 Hz tone in the speaker.

(d) Turn the output of the RF generator all the way down so there is no signal going into the transceiver.

(e) Turn the squelch clockwise until the audio just cuts out.

(f) Increase the signal from the RF signal generator until you can hear the tone in the speaker. The received signal is strong enough to break the squelch.

9. How strong a signal was required to break the squelch?

_____ µV   _____ dBm

(g) Turn the squelch control fully clockwise.

(h) Increase the output of the RF generator until the signal breaks the squelch again.

10. How strong a signal is required to break the squelch when the squelch control is fully on?

_____ µV   _____ dBm

11. What is the range of the squelch control?

    _____ dB

USB/LSB Operation

Technologists often represent upper and lower sidebands with triangles; the point of the triangle represents the lower audio frequency. The following procedure allows us to plot USB and LSB reception on a frequency scale.

(a) Set the transceiver to USB on channel 25.

(b) Set the output of the RF generator to about 0.3 µV.

(c) Adjust the clarifier to its center position.

(d) Adjust the RF generator frequency for the lowest frequency audio tone that you can still hear. Record in the table provided the frequency displayed on the RF generator.

(e) Adjust the RF generator frequency for the highest frequency audio tone that you can still hear. Record in the table provided the frequency displayed on the RF generator.

(f) Set the transceiver to LSB operation and repeat steps (d) and (e).

12. **USB:**

    RF generator frequency = _____ MHz   when lowest frequency audio tone is heard.

    RF generator frequency = _____ MHz   when highest frequency audio tone is heard.

    LSB:

    RF generator frequency = _____ MHz   when lowest frequency audio tone is heard.

    RF generator frequency = _____ MHz   when highest frequency audio tone is heard.

13. Plot triangular sidebands using the frequencies recorded in question 12. Indicate which is the USB operation and which is the LSB operation.

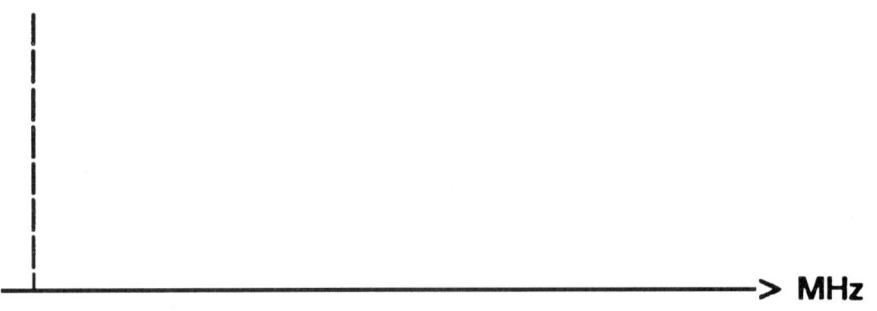

RECEIVER SELECTIVITY

LAB 12

Name: _____   Date: _____

OBJECTIVES:

Upon completion of this lab, you will be able to:

- Interpret correctly the adjacent channel selectivity specification for a receiver
- Measure the adjacent channel selectivity for a receiver

TEST EQUIPMENT:

- A Citizen's Band (CB) radio with the following specifications and to be operated in this lab with the following operating constraints:
    1. The CB radio is designed and manufactured to operate *only* on a low voltage of +13.8 volts dc (nominal) and will be powered during this lab by a low dc voltage of +13.8 volts dc from an adjustable low voltage lab bench dc power supply.
    2. The CB has a 40-channel capability, each channel 10 kHz wide, in the range from 26.965 MHz to 27.405 MHz with a maximum transmitted power in AM mode of 4 watts and 12 watts peak envelope power (PEP) in SSB.
    3. The CB unit is to be operated only in receive mode throughout the lab (a microphone will not be issued). The CB in receive mode when powered with a +13.8 volt dc power supply has a maximum audio output power of 3 watts into an 8 Ω load.
    4. A microphone is not to be part of the lab equipment, so choose a CB that can operate in receive mode when the microphone is not connected to the radio (UNIDEN model PC122XL or equivalent).
- Two RF signal generators
- Three i/p matching pad
- True rms voltmeter
- External 8 Ω speaker—President Model 711-SX or equivalent and audio cabling
- Adjustable low voltage dc power supply (GW model GPC-3020 or equivalent)

Introduction

Selectivity is one of the more important specifications of a receiver. An adjacent channel selectivity specification for a receiver indicates how well a receiver continues to receive a signal to which it is tuned when a strong signal is present on an adjacent channel.

The equipment setup shown in Figure 12-1 illustrates the conceptual setup of equipment needed. Your instructor will provide you with a diagram specific to the particular model of equipment you have in the lab. A microphone will not be part of the lab equipment to avoid accidentally transmitting into the signal generator.

Using the diagram provided by your instructor, connect the test equipment but **do not turn on the power** to the equipment until the instructor has checked your setup.

> *NOTE:* It is dangerous to connect an antenna to a receiver near power lines or in confined spaces such as in a lab. At no time in this lab should there be an antenna hooked to the transceiver.

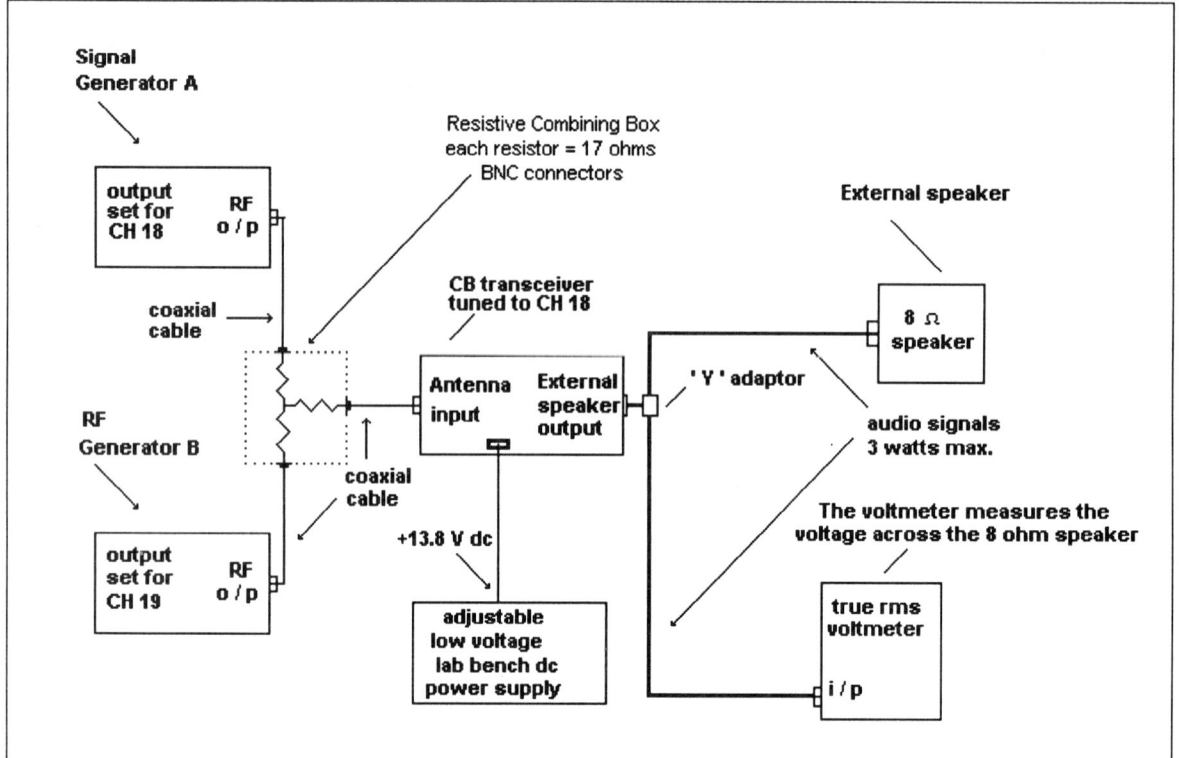

FIGURE 12-1

Instructor's signature: _____

Now turn on the power to your equipment (+13.8 V dc from an adjustable low voltage bench dc power supply).

Procedure

Signal generator A will be the **desired signal**.

- Set to channel 18 or 27.175 MHz.
- Output is amplitude modulated with a modulation frequency of 1000 Hz and a modulation index of 30%.
- Set an output power of about −110 dBm or at an appropriate level to produce an (S+N)/N of 10 dB from the receiver when operating the receiver audio power output (or volume) at half the rated maximum audio output power.

The **receiver** will be tuned to the desired signal.

- Set to channel 18, fully unsquelched.
- With the RF output from signal generator A fine-tuned to give a 10 dB (signal + noise)-to-noise ratio at the receiver audio output, simultaneously adjust the audio power output or volume so that the audio output power is half the rated maximum audio output power of the receiver.

1. What is the output power level in dBm of signal generator A when the receiver has a 10 dB (S+N)/N ratio when operating at half power?

Signal generator B will be the **undesired adjacent channel signal**.

- Set to channel 19 or 27.185 MHz.
- Output is amplitude modulated with 400 Hz with a modulation index of 30%.
- The output level is initially set very low (about −140 dBm) so that it is effectively turned off and is not interfering with the ability of the receiver to receive the desired signal to which it is tuned.

Now increase the output level of generator B upward from −140 dBm in steps of 10 dB at a time.

2. What do you notice happens to the audio quality as the strength of the adjacent channel signal is increased?

You should notice that as the adjacent channel signal strength increases that the quality of the received audio (1000 Hz) deteriorates due to an increase in noise. The ability of the receiver to detect a weak signal to which it is tuned is being degraded by the adjacent channel signal. In effect the receiver has lost sensitivity or is said to be *desensed*.

Imagine that you are operating a transceiver such as a CB or cell phone in your car and receiving a weak signal. A car pulls up next to you at a traffic light and starts to transmit on an adjacent channel. Because of the proximity of this other signal, your receiver will desense.

A quantitative measure of the ability of a receiver to resist desensing is the adjacent channel selectivity specification.

Adjust the output of signal generator B (the adjacent channel signal) until the (S+N)/N ratio from the receiver has deteriorated to 6 dB. Do not forget to adjust the volume control of the receiver so that the receiver is operating at an audio output power setting of half its rated maximum output.

3. What is the output level in dBm of signal generator B when the (S+N)/N ratio has deteriorated to 6 dB?

The difference between the measurement in question 3 and question 1 is the adjacent channel selectivity for your receiver. It tells you how much stronger an adjacent channel signal must be in dB relative to the desired signal before the receiver has been desensitized.

EXAMPLE: Suppose the setting of signal generator A in question 1 were –110 dBm and suppose the setting of signal generator B in question 3 were –50 dBm. In such a case the adjacent channel selectivity would be

Adjacent channel selectivity = (–50 dBm) – (–110 dBm)

$$= 60 \text{ dB}$$

4. What is the measured adjacent channel selectivity for your receiver?

5. If available, locate the specification for your transceiver in the user's manual for your transceiver or on the specifications sheet provided by your instructor.

6. Does your receiver meet the manufacturer's spec?

THE PHASE LOCKED LOOP

LAB 13

Name: _____ Date: _____

OBJECTIVES:

Upon completion of this lab, you will be able to:

- Design and build a phase locked loop circuit using a 565 phase locked loop (PLL) integrated circuit
- Measure the free-running frequency and the lock range of the 565 PLL (phase locked loop) circuit
- Demonstrate the basic operation of a phase locked loop

TEST EQUIPMENT:

- Dual low voltage dc power supply
- Oscilloscope
- Function generator
- PLL integrated circuit (565)
- Material —Capacitors, 330 pF and 470 pF
  —Resistor, 10 kΩ

Introduction

Purpose of Phase Locked Loop Circuits

Phase locked loop circuits, PLLs, are employed extensively in the commercial electronics world. We find them used as FM modulators, FM demodulators, and frequency synthesizers. The PLL is used as a frequency synthesizer in multichannel transceivers, televisions, computers, radios, stereo equipment, and test equipment.

Basic Concepts of PLLs

The block diagram of a phase locked loop is shown in Figure 13-1.

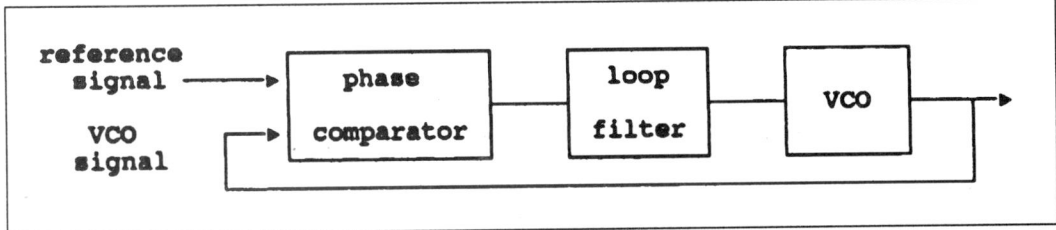

FIGURE 13-1 Block Diagram of a Phase Locked Loop

The phase comparator produces a dc voltage based on the phase difference between the two input signals. f_{ref} is usually a very stable crystal frequency oscillator. f_{vco} is the signal supplied by the voltage controlled oscillator, VCO. When there is zero error voltage supplied to the VCO, it oscillates at its **free-running frequency**. If the free-running frequency and the reference signal are not equal in frequency, a dc error voltage is produced by the phase comparator. This dc error voltage is used to change the frequency of the VCO. Once the error voltage forces the VCO frequency to equal the reference frequency, the VCO locks at this new frequency. Any changes in the frequency of the reference signal will cause a new dc error voltage to be generated at the output of the phase comparator. This new error voltage causes the VCO frequency to follow the reference signal frequency. The difference in frequency between the VCO and the reference voltage must be within a certain range before the VCO will start tracking the frequency of the reference signal. This is called the **capture range** of the PLL. Once the reference frequency is adjusted to fall within the capture range, the VCO will lock onto the reference signal.

Once the VCO has locked onto the reference signal, it will continue to track changes in the frequency of the reference signal until the reference signal frequency moves too far from the free-running frequency of the VCO. The limits over which the VCO will track the reference signal once it has locked onto the reference signal is called the **lock range**. See Figure 13-2.

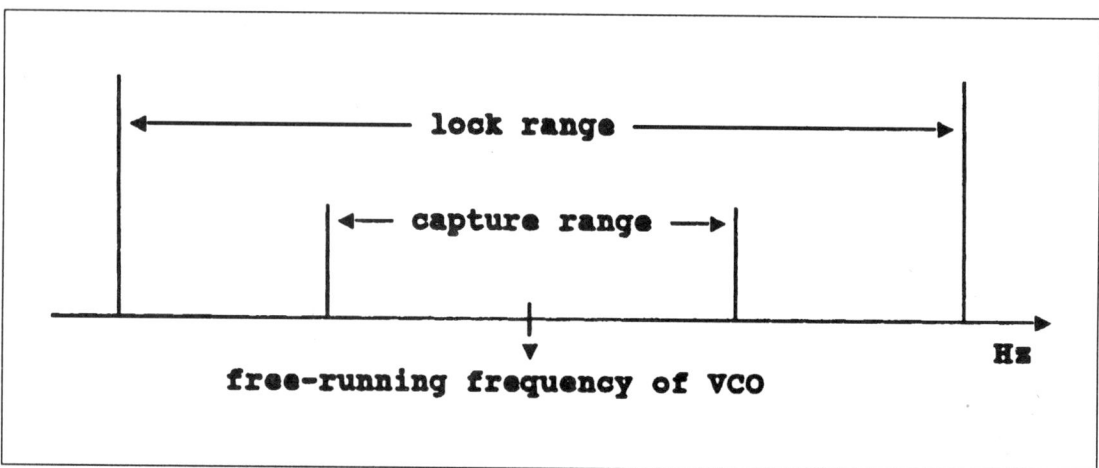

FIGURE 13-2 Lock Range and Capture Range of a PLL

Integrated Circuit Phase Locked Loop (PLL)

Phase locked loops are available in integrated circuit form. An example is the LM565 whose block diagram is shown in Figure 13-3.

Pre-Lab

The principles of operation of a phase locked loop can be readily demonstrated using a phase locked loop circuit such as the 565. The manufacturer of this part provides design equations for such things as the free running frequency and the lock range in terms of the values of resistors and capacitors to be connected external to the part.

The equation for the free running frequency of the 565 is

$$f_o \cong \frac{0.3}{R_o C_o}$$

in which f_o is the free run frequency in Hertz, and R_o and C_o are the values of the resistor in ohms and capacitor in farads to be attached externally to the part.

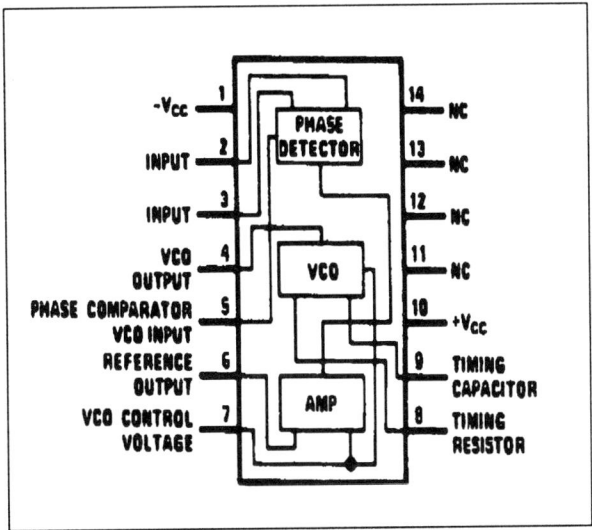

FIGURE 13-3 Block Diagram of the LM565 Phase Locked Loop Circuit (Courtesy of National Semiconductor Corp.)

The equation for the lock range is

$$f_L = 16\, f_o/V_c$$

in which f_L is the lock range in Hertz and V_c is the value in volts of the *total* power supply voltage. (If $V_{cc} = \pm 10$ volts, then $V_c = 20$ in the equation.)

1. Using the above equations, calculate the free run frequency and lock range for a 565 PLL using the following external resistor and capacitor values:

   $R_0 = 10\ k\Omega \qquad C_0 = 470\ pF \qquad V_{cc} = \pm 10$ volts

Procedure

Assemble the phase locked loop circuit as shown in Figure 13-4.

2. Using an oscilloscope, measure the free-run frequency of the VCO in the PLL circuit (at pin 4 or pin 5).

 Answer: **You should find that this frequency is close to the free-running frequency predicted by your calculations in question 1.**

Set a function generator to output a square wave with 3 volts p-p at a frequency equal to the free run frequency calculated in question 1 of this lab.

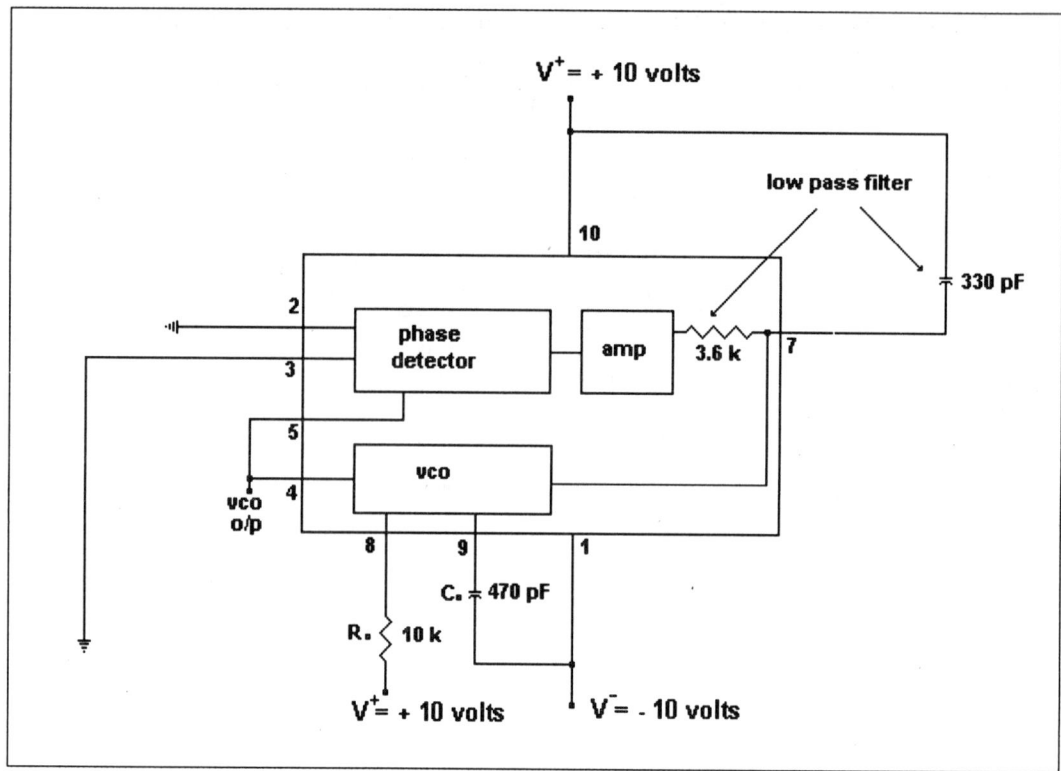

FIGURE 13-4 Using the LM565 PLL Circuit

Connect this signal to the input (pin 2) of the PLL circuit (but do not leave pin 2 grounded!) and monitor the signal on channel 1 of the scope.

Trigger on channel 1.

Connect channel 2 of the scope to monitor the VCO output of the LM565 PLL (pin 4 or 5).

The signal from the function generator should have been captured by the PLL. Vary the frequency of the signal from the function generator to see if the two signals track together. If they are not in lock, the VCO trace on channel 2 will be a blur on the scope display, in which case ask for assistance. If the two signals lock, adjust the frequency of the function generator until the two waveforms are out of phase by 90°. In this state, the frequency of both the VCO and function generator will be the same and very close to the design free-run frequency.

3. Sketch the two waveforms and indicate on the diagram the VCO frequency of the PLL when the phase difference is 90°.

Measure the lock range of the PLL circuit by first increasing the frequency of the function generator above the free run frequency until lock is lost. The frequency of the function generator at which this occurs marks the upper limit of the lock range.

4. What is the upper limit of the lock range?

Now decrease the frequency of the function generator below the free run frequency until lock is again lost. The frequency at which this occurs marks the lower limit of the lock range.

5. What is the lower limit of the lock range?

The difference between the upper limit and the lower limit is the lock range of the PLL demodulator circuit.

6. What is the measured lock range?

7. What was the predicted lock range (question 1)?

The measured and predicted lock range should have compared fairly well.

Inspect the design equations used in question 1 for predicting the lock range and the free run frequency of your PLL demodulator. Note that changing the power supply voltage should change the lock range but not the free run frequency. Decrease the V_{cc} from ±10 volts to ±8 volts.

8. What is the predicted value for the new lock range?

9. What is the measured value of the new lock range?

The equations predict that approximately a 20% increase in the lock range should have resulted due to the 20% decrease in the power supply voltage from 10 volts to 8 volts. Return the power supply voltage to V_{cc} of ±10 volts.

10. From the design equations, what new value of R_o will double the present free run frequency?

Replace R_o with the newly calculated value and measure the free-run frequency and lock range.

11. Now what are the measured values for the lock range and free-run frequency?

12. Do the measured results confirm what you would expect from the design equations?

FM MODULATION AND DEMODULATION

LAB 14

Name: _____ Date: _____

OBJECTIVES:

Upon completion of this lab, you will be able to:

- Construct a frequency demodulator circuit using a phase locked loop (PLL) integrated circuit
- Measure the free-running frequency and the lock range of the PLL (phase locked loop) demodulator circuit
- Generate a frequency modulated signal
- Use the PLL to demodulate an FM signal

TEST EQUIPMENT:

- Dual low voltage dc power supply
- Dual-trace oscilloscope
- Function generator
- PLL integrated circuit 565
- VCO integrated circuit 566

Frequency Demodulation Using a Phase Locked Loop (PLL)

FM demodulation can be achieved directly, using a phase locked loop circuit such as shown in Figure 14-1. The carrier frequency of the FM signal is centered in the capture range of the phase locked loop causing the PLL to lock onto a frequency modulated signal. As the input frequency varies, such as would occur with a frequency modulated carrier, the output from the phase locked loop filter will vary in amplitude proportional to the frequency deviations of the input signal and at the same rate at which the deviations occur. In effect the FM signal will be demodulated by the PLL.

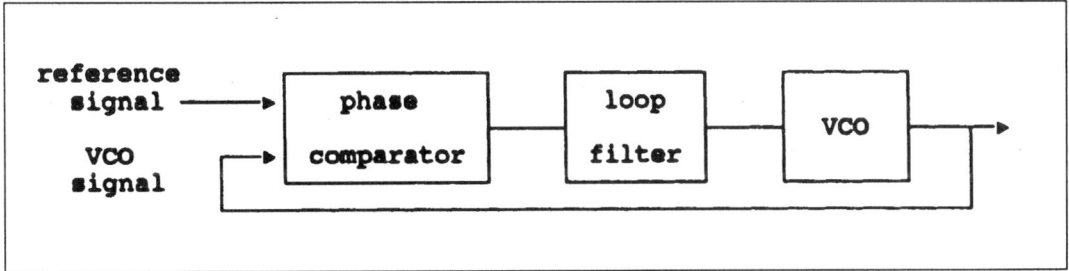

FIGURE 14-1 Block Diagram of a Phase Locked Loop Circuit

Phase Locked Loop Operation

The part we will use for this lab is the 565, which has the basic components of a phase locked loop in an integrated circuit package. The equations for the free-running frequency and lock range of the 565 are provided below and refer to Figure 14-2.

The equation for the free running frequency of the 565 is

$$f_o \cong 0.3/R_o C_o$$

where R_o and C_o are the values of the resistor in ohms and the capacitor in farads to be attached externally to the part, and f_o is the free run frequency in Hertz.

The equation for the lock range is

$$f_H = 16 f_o / V_c$$

where V_c is the value in volts of the total power supply voltage (if $V_{cc} = \pm 10$ volts, then $V_c = 20$ volts in the equation) and f_H is the lock range in Hertz.

1. Use these equations to calculate the free run frequency and lock range for a 565 PLL using the following external resistor and capacitor values.

   $R_0 = 10$ kΩ $C_0 = 470$ pF $V_{cc} = \pm 10$ volts

Assemble the phase locked loop circuit as shown in Figure 14-2.

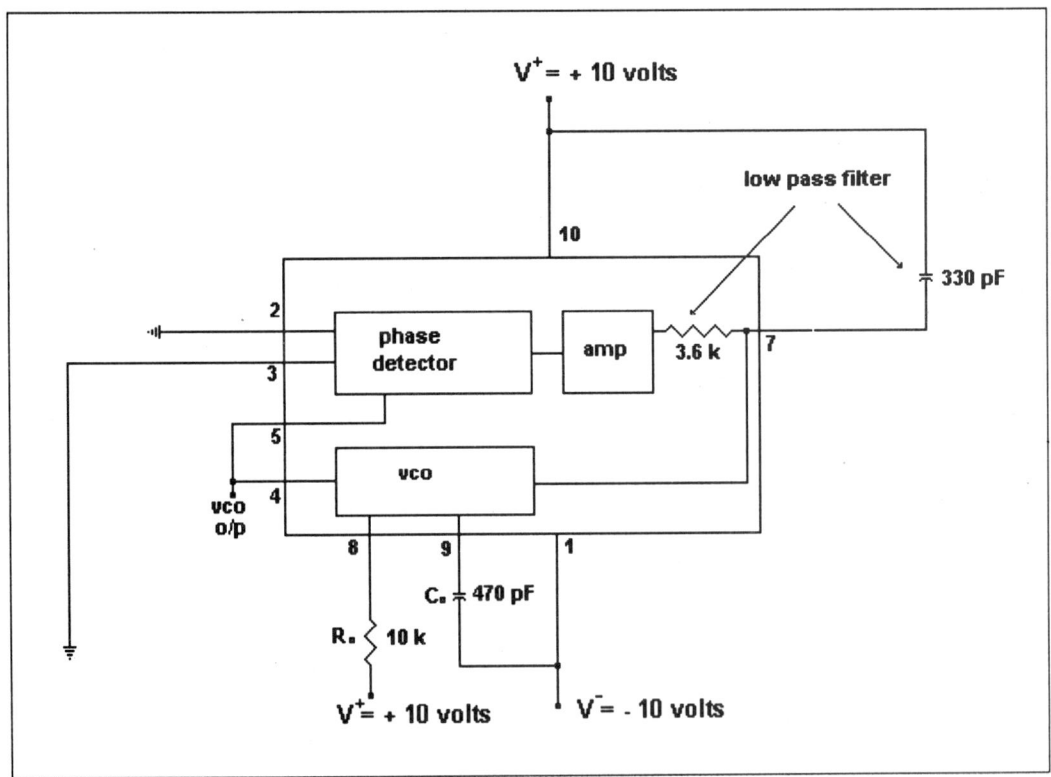

FIGURE 14-2 Using the 565 PLL Circuit

2. Measure the frequency of the VCO in the PLL circuit (at pin 4 or pin 5).

You should find that this frequency is close to the free-running frequency predicted by your calculations in question 1. If the VCO frequency of the PLL is much different than the expected carrier frequency, the PLL demodulator circuit will not be able to lock onto the carrier signal.

Generate such a carrier by adjusting the function generator to output a square wave with 3 volts p-p at a frequency equal to the free run frequency calculated in question 1 of this lab.

Connect this signal to the input (pin 2) of the PLL circuit and monitor the signal on channel 1 of the scope. Trigger on channel 1.

Connect channel 2 of the scope to monitor the VCO output of the LM565 PLL (pin 4 or 5).

The signal from the function generator should have been captured by the PLL. (If the PLL has not gone into lock mode the VCO output on the scope will be blurred.) Vary the frequency of the signal from the function generator and note that the two signals track or remain "locked" together on the oscilloscope display. As the frequency of the function generator is varied, so does the VCO frequency. Adjust the frequency of the function generator until the two waveforms are out of phase by 90°.

When the two signals are out of phase by 90°, the frequency of the VCO will be close to the free run frequency calculated previously.

3. What is the measured VCO frequency when the two signals are out of phase by 90°?

The lock range of a PLL is a critical factor in demodulating an FM signal. Make a rough estimate of the lock range of the PLL circuit by first increasing the frequency of the function generator above the free run frequency until lock is lost, and then decreasing the frequency of the generator below free run until lock is lost. The difference between these two frequencies is the lock range of the PLL demodulator.

4. What is the measured lock range?

Generating an FM signal

To generate a frequency modulated signal, we will use a voltage controlled oscillator (VCO). As the name implies, the voltage on the input of the VCO will determine the frequency of the VCO output. An integrated circuit that fulfills this function is the LM566, and its block diagram is shown in Figure 14-3. In this diagram, the output on pin 3 is a square wave whose frequency is determined by the control voltage V_c on pin 5.

Assemble the circuit shown below in Figure 14-4. Using a digital multimeter, adjust the potentiometer to provide an input voltage of 8.0 volts.

Measure the output frequency of the 566 VCO at pin 3.

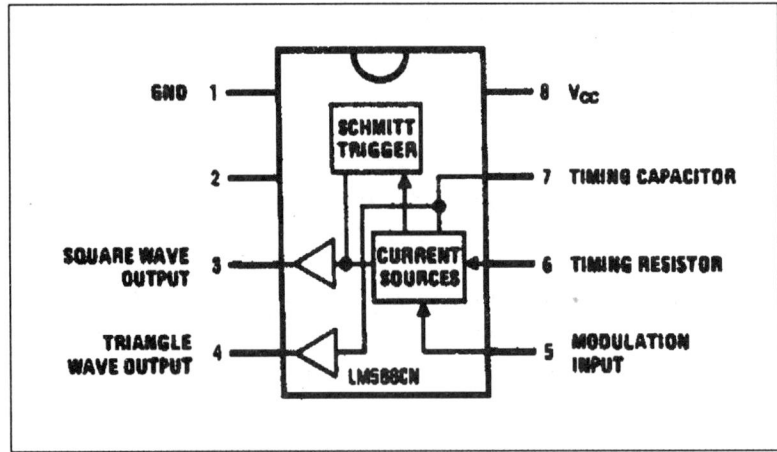

FIGURE 14-3 Block Diagram of the LM566 (Courtesy of National Semiconductor Corp.)

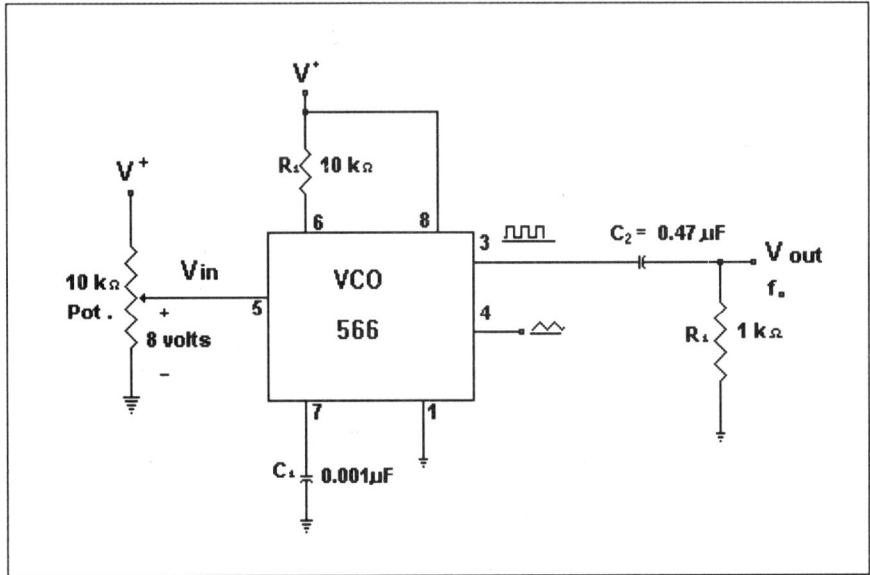

FIGURE 14-4 Voltage Controlled Oscillator

5. What was the measured output frequency?

The theoretical output frequency for the LM566 VCO is given by the following equation:

$$f_o = \frac{2.4}{R_1 C_1}\left(1 - \frac{V_5}{V^+}\right)$$

where V_5 is the voltage between pin 5 and pin 1.

Using the values from the 566 VCO circuit you have tested, calculate the frequency predicted by the design equation.

6. What is the theoretical VCO frequency?

7. How well (or poorly) did the predicted value compare with the measured value?

Using a DMM, adjust the potentiometer down to +7 volts and up to +9 volts and back to +8 volts. Observe the change in the VCO output frequency.

8. What was the VCO frequency at +7 volts?

9. What was the VCO frequency at +9 volts?

Remember to return the potentiometer setting to +8 volts. Connect the circuit shown in Figure 14-5. Note that a coupling capacitor has been added to isolate the function generator from the DC bias at the control input of the 566.

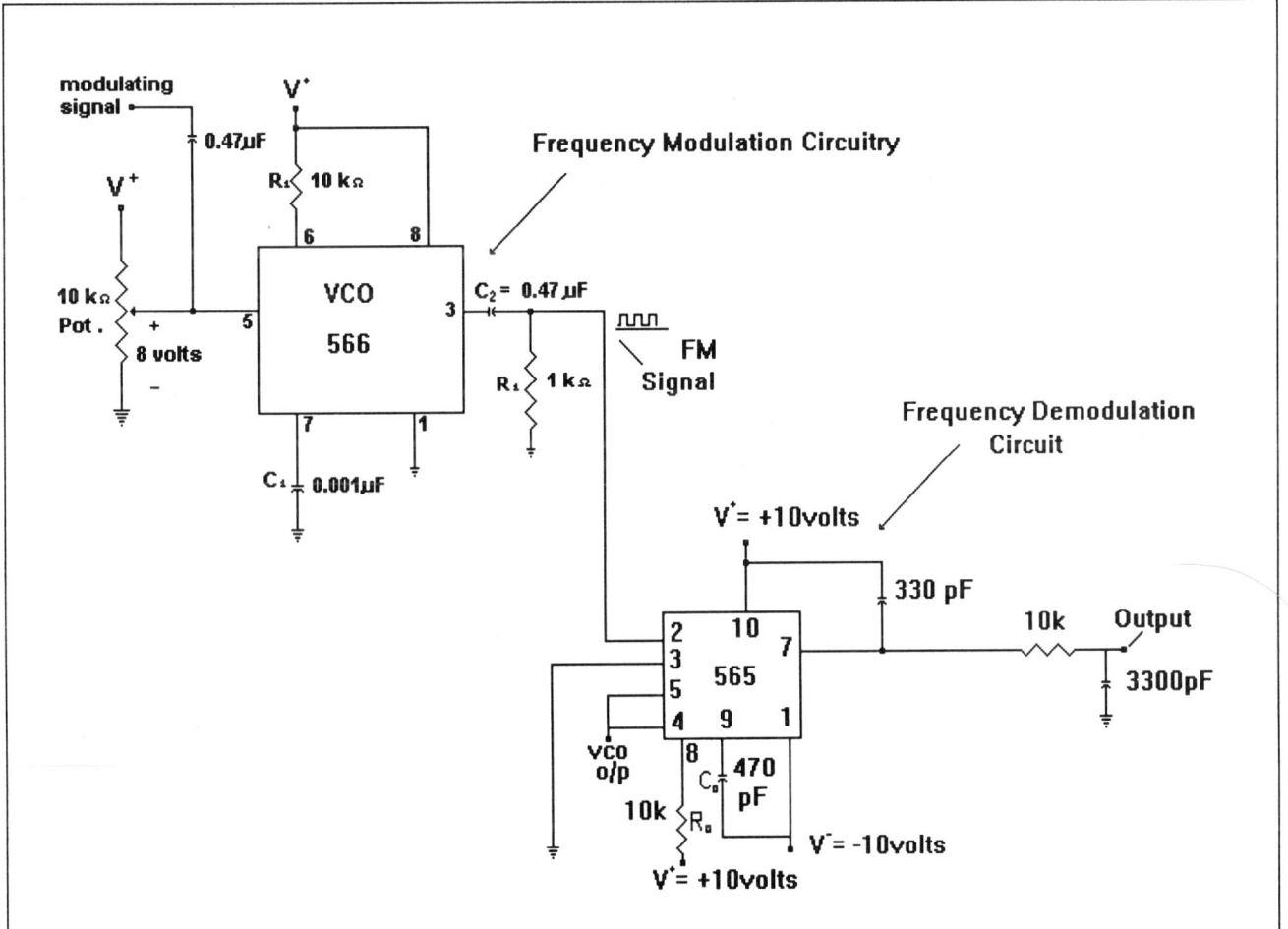

FIGURE 14-5 Frequency Modulation/Demodulation

Adjust the function generator to provide a 1 kHz 2 V p-p sine wave at the input of the LM566 VCO. Monitor this signal with channel 1 of the oscilloscope.

Connect channel 2 of the oscilloscope to the demodulated and filtered output of the PLL at the test point marked Output in Figure 14-5.

The signals on channel 1 and channel 2 should both have the same waveshape and frequency.

> *NOTE:* It is possible that the modulating signal amplitude is too large and is causing a total deviation on the carrier which exceeds the lock range of the PLL. If this is the case, you should notice that the PLL loses lock periodically during the modulating signal's cycle, causing the demodulated output on channel 2 to be distorted. If this occurs, turn down the amplitude of the modulating signal until the PLL can track the frequency deviations of the carrier without losing lock.

10. Sketch the input and output voltages.

Vary the frequency and waveshape of the modulating signal. The demodulated output should track these changes.

Increase the amplitude of the modulating signal and see if the PLL demodulator loses lock when the frequency deviations on the modulated carrier exceed the lock range of the PLL demodulator.

11. Comment on the importance of lock range in determining the ability of a PLL to function as a demodulator.

FM RECEIVER SENSITIVITY

LAB 15

Name: _____ Date: _____

OBJECTIVES:

Upon completion of this lab, you will be able to:

- Measure the quieting sensitivity of an FM receiver
- Describe the SINAD sensitivity test for an FM receiver

TEST EQUIPMENT:

- A VHF FM scanner with the following specifications, to be used in this lab only for FM receiver tests with the following operating constraints:
  1. The scanner (FM receiver) is designed and manufactured to be operated *only* from a low voltage of +13.8 volts dc (nominal) and is to be powered during this lab by a low dc voltage of +13.8 volts dc from an adjustable low voltage lab bench dc power supply.
  2. The FM receiver has an external 8 Ω speaker jack, squelch control, and BNC or standard connector at the antenna input.
  3. When powered with +13.8 volts dc, the receiver's maximum audio output power must not exceed 3 watts into an external 8 Ω external speaker load. Also, please see the note below.
- True RMS voltmeter (1), RF signal generator (1), cabling
- 8 ohm external speaker box—President Model 711-SX or equivalent
- Low voltage adjustable dc power supply (GW model GPC-3020 or equivalent)

> *NOTE TO INSTRUCTOR:* It is the purpose of this lab to use a scanner in order to have a low audio output power FM receiver powered from a low voltage dc power supply for radio system tests in an educational setting. However, the possession, installation, or use of a scanner may be prohibited, regulated, or require a permit in certain countries, states, provinces, cities, and/or local jurisdictions. Before proceeding, check with local law enforcement officials who should be able to provide you with information with respect to the pertinent laws and regulations.

FM Receiver Sensitivity Measurements

Pre-Lab

There are a number of questions that must be answered before coming to the lab. Read through the lab material and answer all possible questions.

Procedure

The diagram in Figure 15-1 is a conceptual sketch of the equipment setup in this lab. Your instructor will provide you with a detailed diagram of the required equipment setup as well as a handout detailing the specifications of the specific scanner you will be using to do the FM receiver tests. Use the diagram provided by your instructor to connect the equipment but **do not turn on the power** to any of the equipment until your instructor has checked your setup.

> *NOTE:* It is dangerous to connect an antenna to a receiver near power lines or in confined spaces such as in a lab. At no time in this lab should there be an antenna hooked to the receiver.

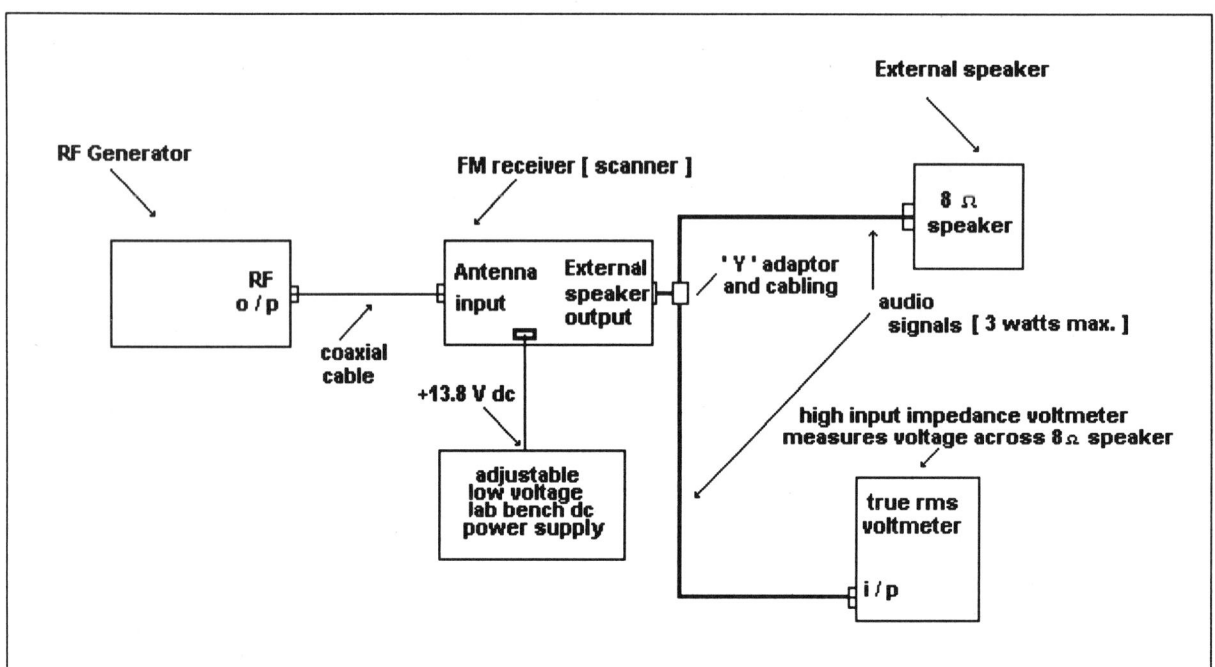

FIGURE 15-1

 Instructor's signature: _____

Now turn on the power to the equipment. (+13.8 volts dc from an adjustable low voltage lab bench dc power supply).

FM Thresholding

Before performing this test, answer the following question. (Use your text as a reference.)

1. Explain the quieting of an FM receiver in the presence of a strong enough **unmodulated** carrier. Make reference in your answer to FM thresholding.

Refer to the instructions provided for the particular scanner (FM receiver) you are using and set the receive frequency on your scanner for a value set by your instructor in the pre-lab briefing. A typical value might be 145.00 MHz.

Set the RF generator to the receiver frequency with the generator output power set to the minimum possible output level (effectively the input signal to the receiver has been removed) and with no modulation.

Set the squelch control on the receiver fully counterclockwise and adjust the volume control until the output noise is loud enough for comfort.

Now increase the output level of the generator. As you do this, the audio noise level on the receiver output will go down markedly. The receiver is "quieting" or the signal has "captured" the receiver from the noise.

20 dB Quieting Sensitivity

A test that measures the quieting of an FM receiver quantitatively is the 20 dB quieting sensitivity.

A receiver's quieting sensitivity is tested when the audio output power from the receiver is operating at 25% of the rated maximum output. (The 25% rating is an agreed-upon standard). As an example, suppose the rated maximum audio output power of the receiver is 2 watts. This would make 1/4 power to be 0.5 watt.

It is possible to set the receiver to operate at 1/4 power by adjusting the receiver volume control until the rms voltage across the external 8 ohm speaker load is 2.00 volts.

Since 2 volts across 8 ohms dissipates 0.5 watts, the receiver would be operating at 1/4 power.

2. From the specification sheets, find the maximum rated output power for your receiver and calculate the rms voltage needed across an external 8 Ω load to output audio power at 25% of the maximum rating.

Adjust the volume control on the receiver to set this voltage across the 8 ohm external speaker load.

To measure the quieting sensitivity spec, proceed as follows:
- Set the RF generator to the receiver frequency with minimum output level (effectively the input signal to the receiver has been removed) and **no modulation**.

- Adjust the squelch control until the receiver is fully on. This will be indicated by the presence of noise on the output of the receiver.
- If you have not already done so, adjust the volume control so that the noise power out is 25% of rated maximum output as indicated by the voltmeter reading calculated in question 2.
- Increase the RF generator output with **no modulation** until the receiver audio noise decreases by 20 dB (as the generator output goes up, the voltmeter reading will go down). At this point the generator output level is the quieting sensitivity.

3. What should the voltmeter reading be when the noise power decreases by 20 dB? Show your calculations.

4. What is the measured quieting sensitivity as indicated by the RF generator output level in microvolts?

5. Does your transceiver specify a value for the quieting sensitivity of your receiver? If so, state the value and indicate if the receiver met the stated specification.

SINAD Sensitivity

A common sensitivity test for FM receivers is the SINAD test.

6. Using a block diagram, explain what is involved in making a SINAD test. Reference your text or other suitable source.

The specifications provided with your transceiver should state a SINAD value for your receiver.

7. What does the specification sheet for your transceiver say is the SINAD for your receiver?

Squelch Control

The squelch control on a receiver allows muting of the annoying audio noise present in the absence of a signal at the receiver input.

Squelch works by turning off the audio stage if the received signal strength drops below a certain level. This level is set by the squelch control setting, which is adjustable.

The squelch control can be adjusted to mute the receiver for a range of received signal strengths. In this way the operator of a receiver can avoid listening to any signal below a certain strength. In a sense the squelch modifies the sensitivity of a receiver. (The signal that can break squelch sets the level for the weakest signal that can be received.)

Locate the squelch control on the receiver. Usually the control is located on the front of the receiver so that it is easily accessible. Consult the diagram provided by your instructor.

Make the following investigation of the squelch control, using the same equipment setup as before:

- Adjust the frequency of the RF generator to the receiver frequency and set the deviation to ±3 kHz with a modulation frequency of 1000 Hz. Set the output level of the generator to 1 mV.
- Adjust the volume control so that the audio output power is the rated output power of the receiver. For example, if the rated maximum output audio power is 2 watts, the rms voltmeter should measure 4 volts across the external 8 ohm speaker load.

8. What is the rms voltage that is needed across the 8 ohm external speaker for the receiver you are testing to be operating at the rated maximum audio output power?

- Now set the RF generator for its lowest possible output. The intent here is to remove the received signal from the receiver.
- Adjust the squelch control on the receiver until the audio stage is turned off (no more noise!).
- Slowly increase the output of the signal generator until the signal "breaks squelch" and the audio comes back on.

When the signal breaks squelch, note the signal generator output level in μvolts at which this happens. Then repeat the above procedure, but this time set the squelch control for its maximum position. Again note the generator output level that breaks squelch.

9. What was the smallest signal that can be squelched out?

10. What is the largest?

FM CAPTURE EFFECT

LAB 16

Name: _____   Date: _____

OBJECTIVES:

Upon completion of this lab, you will be able to:

- Demonstrate the desensing of an FM receiver due to strong signals on adjacent channels
- Demonstrate interference due to third odd-order IMDs
- Demonstrate the capture effect of an FM receiver

TEST EQUIPMENT:

- A VHF FM scanner with the following specifications, to be used in this lab only for FM receiver tests with the following operating constraints:
    1. The scanner (FM receiver) is designed and manufactured to be operated *only* from a low voltage of +13.8 volts dc and is to be powered during this lab by a low dc voltage of +13.8 volts from an adjustable low voltage lab bench dc power supply.
    2. The FM receiver has an external 8 Ω speaker jack, squelch control, and a BNC or other standard connector at the antenna input.
    3. When powered with +13.8 volts dc, the receiver's maximum audio output power must not exceed 3 watts into an external 8 Ω external speaker load. Also, please see the note below.
- Three RF generators, cabling
- One 3-input 1-output 10 dB resistive combining pad
- Adjustable low voltage dc power supply (GW Model GPC-3020 or equivalent)

> **NOTE TO INSTRUCTOR:** It is the purpose of this lab to use a scanner in order to have a low audio output FM receiver powered from a low voltage dc power supply for radio system tests in an educational setting. However, the possession, installation, or use of a scanner may be prohibited, regulated, or require a permit in certain countries, states, provinces, cities, and/or local jurisdictions. Before proceeding, check with local law enforcement officials who should be able to provide you with information with respect to the pertinent laws and regulations.

Introduction

Intermodulation interference is produced when two strong signals, existing at frequencies other than the frequency to which the receiver is tuned, mix together to produce intermodulation distortion products (IMDs) that lie in the pass band of the receiver. The IMDs are generated when the RF amp at the front end of a receiver is driven into nonlinear action by the presence of very strong signals at the receiver input. The receiver has filtering tuned to reject signals on adjacent channels. However, when these adjacent channel signals are strong enough, the residual signals after attenuation by the front end filtering may still be large enough to drive the receiver's front end RF amplifier into nonlinear operation. If the undesired signals are strong enough, the intermodulation products produced may well desensitize the receiver. Receivers are designed to provide a rejection of 50 to 60 dB. This means the undesired signals must be 50 to 60 dB stronger than the desired signal to produce significant interference.

In some cases the undesired signals may be so strong that the resulting intermodulation product may be stronger than the desired signal. The net result is the desired signal will be lost and the intermodulation product will be received. This is an example of the FM *capture effect*. Any FM receiver will lock onto the strongest signal so long as it is a few dB greater than any other signal. The other weaker signals produce no visible interference. As soon as a weaker undesired signal becomes a few dB stronger than the desired signal, the receiver locks onto the undesired signal and the desired signal is lost. This demonstration is designed to illustrate the FM capture effect.

The increase in popularity of FM communications has resulted in a battle for a good location to place the system's antenna. With FM communications, it is important to get the antenna as high as possible to increase the range of transmission. This has led to a congestion of antennas on tall buildings and hills situated close to metropolitan areas. (Next time you drive by downtown, check how many antennas are mounted on the taller buildings). This close proximity of antennas provides many very strong signals which could result in intermodulation interference. The problem is getting so complicated that computers are often used to predict whether two frequencies with antennas close together will interfere with another station that is nearby.

How Intermodulation Interference Is Produced

Assume you are in your car receiving a signal at 145.00 MHz. The channel spacing is 30 kHz for this band of frequencies. Two friends drive up in their cars with similar radios. One friend is operating at the next channel up, and the other friend is two channels up from you. We have the following situation:

f_1 = Your frequency = 145.00 MHz
f_2 = Friend #1's frequency = 145.00 MHz + 30 kHz
f_3 = Friend #2's frequency = 145.00 MHz + 60 kHz

Intermodulation interference occurs when we have $2 \times f_2 - f_3$. The result is shown below.

$2f_2 - f_3$ = 2(145.00 MHz + 30 kHz) − (145.00 MHz + 60 kHz)

= 290.00 MHz + 60 kHz − 145.00 MHz − 60 kHz

= 145.00 MHz

The 145 MHz result is an example of what is called a *third odd-order IMD product*. Notice too that this intermodulation product is the same frequency as the channel you are receiving on! Since the two mobile radios are in cars right next to you, the signals at your receiver's input will be very strong. This can result in your receiver being desensitized so you will not be able to hear a weak signal on the frequency you are tuned to. Even worse—if the two signals are strong enough, you will pick up the nearby mobiles instead of the channel you want to receive.

Procedure

The equipment setup shown in Figure 16-1 is the conceptual sketch of the equipment for this lab. Your instructor will provide you with a more detailed diagram of the exact equipment setup to be used. Using the diagram provided by your instructor, connect the equipment but **do not turn on the power** to any of the equipment until your instructor has checked your setup.

> *NOTE:* It is dangerous to connect an antenna to a receiver near power lines or in confined spaces such as in a lab. At no time in this lab should there be an antenna hooked to the receiver.

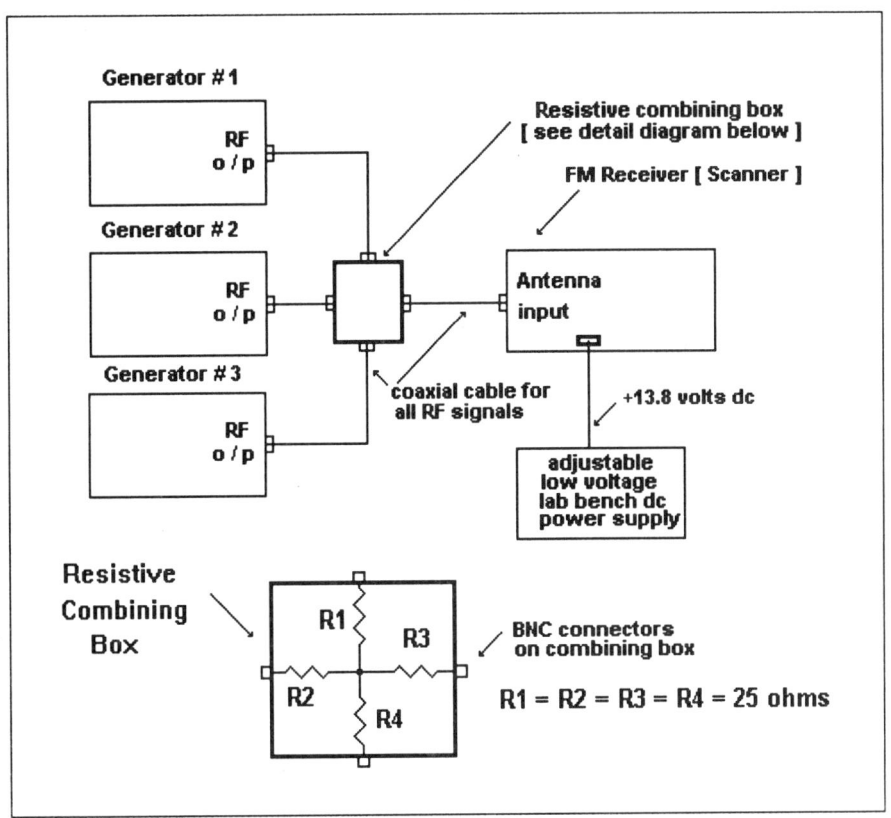

FIGURE 16-1 Intermodulation Interference Demonstration

Instructor's signature: _____

Now turn on the power to the equipment.

Set signal generators 2 and 3 for a very small output of −120 dBm.

Adjust the frequency of signal generator 1 for 147.33 MHz and modulate the carrier with ±3 kHz deviation using a modulation frequency of 1000 Hz.

> *NOTE:* Choosing the frequency of 147.33 MHz is arbitrary and could be changed to a more convenient value if desired. What is important in this lab is that the frequencies of the generators be separated from one another by the channel spacing of the receiver.

Set the receiver to receive at 147.33 MHz.

Consult the receiver's spec sheets for the receiver's SINAD sensitivity. The SINAD spec is almost always quoted in μvolts, but it can be converted to dBm by remembering that the input of your receiver is 50 ohms.

Adjust the output level of signal generator 1 so that the signal from this generator at the antenna input is the SINAD value. (Do not forget to account for a signal loss of 10 dB in the 3-to-1 resistive combining network used in Figure 16-1.)

1. What is the setting of the signal generator in dBm?

　　Set the squelch control fully CCW and turn up the audio control (volume). Since the signal at the antenna input of the receiver is the SINAD value, the noise component of the output audio signal should be very audible. If it is not, turn the generator output level down until this is the case. Since the measurement we are making is only to demonstrate an effect, we will not be too precise about settings on the generator.

　　Set signal generator 2 for a frequency of 147.33 MHz + 30 kHz = 145.36 MHz. This is one channel up from the desired channel. Do not use any modulation on this signal.

　　Set signal generator 3 for a frequency of 147.33 MHz + 60 kHz = 145.39 MHz. This is two channels up from the desired channel. FM modulate the carrier with ±3 kHz of deviation using a 400 Hz sine wave.

　　Simultaneously increase the output of generators 2 and 3 until the signal-to-noise ratio from the receiver declines noticeably. (The noise which was very audible before has now become very, very audible.)

　　As you have observed, the receiver has lost sensitivity due to strong signals on adjacent channels.

　　Continue to simultaneously increase the output of generators 2 and 3 until the 1000 Hz tone disappears and the 400 Hz tone becomes prominent. (When this occurs we say the receiver has been "captured" by the new signal—in this case, an IMD!)

2. What is the output level of generator 2 or 3 in dBm when this occurs?

3. How much stronger in dB are the signals on either of the adjacent channels compared to generator 1 when the IMD captures the receiver?

4. List the two objectives of this demonstration.

5. Explain the role the capture effect plays in allowing cell phones to re-use frequencies. (Refer to your text for this.)

6. An example of another IMD product would be $3f_2 - 2f_3$. This is called a fifth odd-order IMD. If f_2 were two channels (60 kHz) above f_1 and f_3 were three channels (90 kHz) above f_1 (which is at 145 MHz as before) where would this IMD product lie?

7. Repeat the last question but make f_1 = 945 MHz. Does anything change?

8. You are operating at 144 MHz with a channel spacing of 30 kHz. What are the two lower-channel frequencies that could combine to produce a third odd-order IMD at 144 MHz?

DATA TERMINAL I/O TROUBLESHOOTING

LAB 17

Name: _____ Date: _____

OBJECTIVES:

Upon completion of this lab, you will be able to:

- Use a breakout box to troubleshoot a serial i/o port
- Demonstrate the ability to change the communications settings on a PC or data terminal

TEST EQUIPMENT:

- Configurable data terminal (Qume 101 or equivalent) and manufacturer's instruction manual for changing the communications settings of the data terminal

  or

- Personal Computer in terminal mode of a communications software package such as PROCOMM PLUS™ and the associated software manual
- RS-232 Tri-State Breakout Box and I/O Tester (DataCom Technologies Model 600 or equivalent) and manufacturer's user manual for breakout box signal interpretation
- RS-232 cable, oscilloscope

Prerequisites

To perform this lab, the student should have received instruction on the following topics:

- Asynchronous transmission
- RS-232 serial communications standard
- Differences between DTEs and DCEs
- ASCII character set
- Half duplex and full duplex operation
- Breakout box (BOB)
- Data terminal/PC communications settings
- Parity

Equipment Setup

Figure 17-1 is a conceptual sketch of the lab equipment setup. Your instructor will provide you with a more detailed diagram specific to the particular equipment you are working with. With the **power to all equipment turned off** (and which is to remain off until the equipment setup has been checked by your instructor and you have received a check off signature), hook up the equipment as shown in the instructor's detailed diagram. This diagram will show the equipment setup including how and where to:

1. Connect the RS-232 cable to the serial RS-232 port on the back of the data terminal or computer
2. Connect the other end of the RS-232 cable to the **DTE** side of the breakout box (BOB)

Have the instructor check your equipment setup.

 Instructor's signature: _____

Now turn on the power to the equipment (data terminal/PC, breakout box).

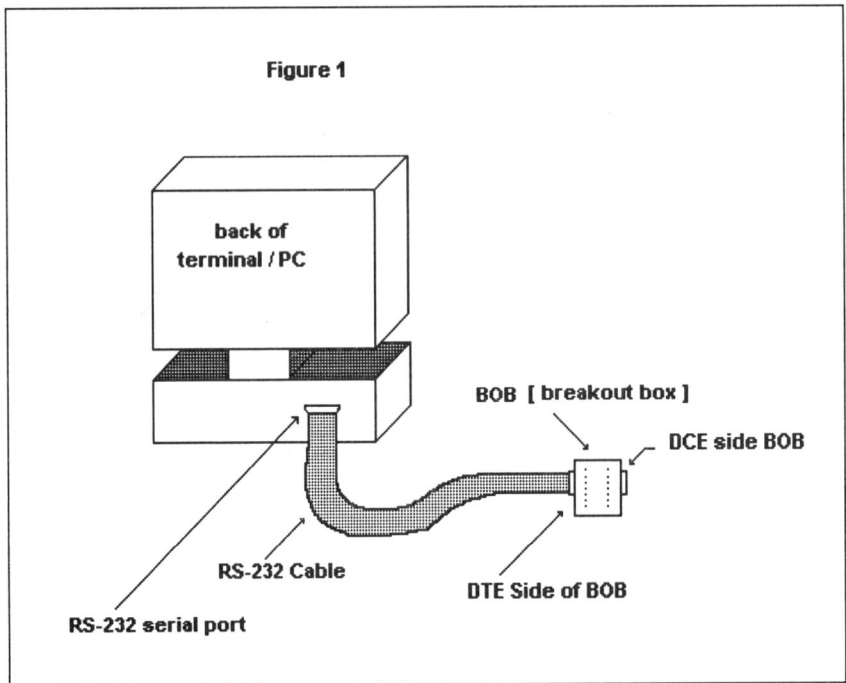

FIGURE 17-1 Data Terminal Connections

Data Terminal Settings

When data is transferred between two computers, the communications settings on both machines have to be the same.

For instance, one computer may require 7-bit characters with no parity while another computer might use 8-bit characters with odd parity. To have a successful communication link, the settings would have to be changed to agree with one another. For this reason it is important to know how to manipulate the communications settings on your data terminal or PC.

If you are using a configurable data terminal, the communications settings on a modern data terminal can be easily changed by keystrokes guided by a pop-up menu. Your instructor will provide you with the manual for the particular model of data terminal at your lab station.

If you have a PC at your work station, the settings on a personal computer can be changed by keystrokes or mouse clicks when the machine is placed into the terminal mode of a communications software package such as PROCOMM PLUS™ or other popular software package.

Study the instructions provided for your workstation and answer the following questions.

1. What keys must be pressed or process followed to put your terminal into a mode where its status can be set or altered?

2. What key sequence must be pressed or process followed to save the settings so they will appear the next time the terminal is turned on?

3. Why is it important to be able to change the settings on a terminal and then save them as discussed in question 2?

Consult the instruction sheets for your terminal/PC on how to change the status of your terminal and adjust your terminal/PC for 1200 bits per second with line feed on. Do so, and then answer the following questions.

4. (a) What are the possible baud rates?

   (b) What baud rate is your terminal set for?

5. (a) Which transmission mode must the terminal be in (HDX or FDX) if the characters are to appear on the screen regardless of whether there is any equipment on the other end?

(b) What transmission mode is your data terminal in?

6. Describe the data format your data terminal is using:

   Data bits,   Bit 8,
   (8/7) _____   (1/0) _____

   Parity,   Parity bit,
   (on/off) _____   (even/odd) _____

   Stop bits,
   (1/2) _____

7. Including the start-bit, how many bits make up each character that is transmitted?

Loop Back Test

With the equipment set up as explained previously, make note of which LEDs on the breakout box are lit.

As you know the breakout box (BOB) is a troubleshooting tool for checking the RS-232 signals sent from the serial RS-232 port of a computer or terminal. Most BOBs support two color LED display, although some are available with a simple light on/light off capability.

Your instructor will provide you with a copy of the signal description writeup and pinout diagram that came with the particular breakout box at your workstation.

8. Refer to the Breakout Box handouts supplied by your instructor or class notes to answer the following:

   (a) What does a red LED indicate?

   (b) What does a green LED indicate?

9. Which LEDs on your BOB have:

   (a) red LEDs lit? _____

   (b) green LEDs lit? _____

Set all the DIP switches on the breakout box to the open position. This means the signals from the data terminal will come this far and no farther.

The DTE transmits data on pin 2. Press any key and observe the LED connected to pin 2.

10. What happens to the LED connected to pin 2 when a key is pressed? Explain why this happens.

If the terminal is not set for full duplex (FDX), set the status of the terminal to FDX at this time. Now press a key.

Notice nothing shows up on the screen when you press a key. The reason is because the DTE is in the FDX (full duplex) mode. In the HDX (half duplex) mode the DTE sends the character to the screen as well as out the serial port. In FDX, the DTE sends the character only to the serial port and not to the screen.

To prove this point, place a jumper between pin 2 (transmit data) and pin 3 (receive data) on the BOB. Now press a key. The character is sent out on pin 2 and returns on pin 3. It is displayed on the screen as received data.

Another way to prove that data is sent to the screen in HDX but not in FDX is to reconfigure the DTE for HDX. Remove the jumper from pins 2 and 3 on the BOB. Press a key just to refresh the idea that the DTE does not send the character to the screen in FDX mode. Reconfigure the DTE for HDX mode. Change the transmission mode to HDX. Now try pressing a key. The data is displayed on the screen.

Place the jumper between pins 2 and 3. Try pressing a key.

11. What appears on the screen when you press a key while the DTE is in the HDX mode and there is a jumper between the transmit data line and receive data line? Explain why this happens.

Set the DTE back to the FDX mode. The loop back test consists of sending a signal out on pin 2 (the transmit data line), looping it back on pin 3 (the receive data line), and observing the results on the screen while in the FDX mode. If all the connections are okay and the equipment is working correctly, you will receive the character on the screen.

12. If you do not receive the transmitted character on the screen, how can you determine whether the problem is the cable or the DTE?

ASYNCHRONOUS COMMUNICATIONS

LAB 18

Name: _____   Date: _____

OBJECTIVES:

Upon completion of this lab, you will be able to:

- Demonstrate the ability to predict the asynchronous pattern for a given ASCII character when supplied with the number of bits per character, the parity, and number of stop bits
- Use an oscilloscope to monitor the asynchronous data flow from the serial port of a terminal

TEST EQUIPMENT:

- Configurable Data terminal (Qume 101 or equivalent) and manufacturer's instruction manual for changing the communications settings of the data terminal.

  or

- Personal Computer in terminal mode of a communications software package such as PROCOMM PLUS™ and the associated software manual
- RS-232 Tri-State Breakout Box and I/O Tester (DataCom Technologies Model 600 or equivalent) and manufacturer's user manual for breakout box signal interpretation
- RS-232 cable, oscilloscope

Prerequisite

To perform this lab, the student should have completed the previous lab on configuring and troubleshooting a data terminal.

Equipment Setup

Figure 18-1 is a conceptual sketch of the lab equipment setup. Your instructor will provide you with a more detailed diagram specific to the particular equipment you are working with. With the **power to all equipment turned off** (and which will remain off until the instructor has checked your setup and given you a check off signature), hook up the equipment as shown in the instructor's diagram. This diagram will show the equipment setup including where and how to

1. Connect the RS-232 cable to the serial RS-232 port on the back of the data terminal or computer
2. Connect the other end of the RS-232 cable to the **DTE** side of the breakout box (BOB)

Have the instructor check your equipment setup.

Instructor's signature: _____

Now turn on the power to the equipment (Data terminal/PC breakout box).

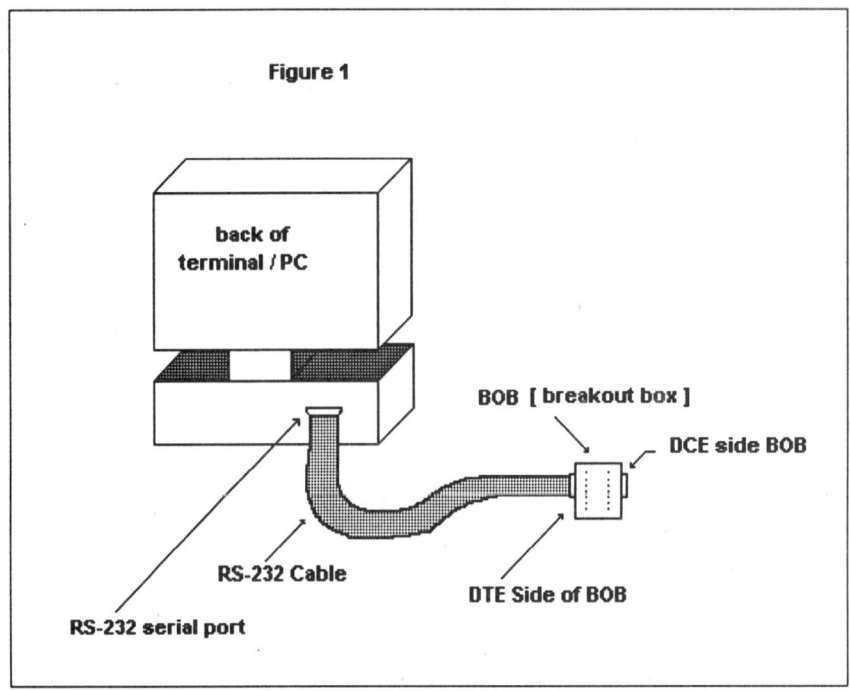

FIGURE 18-1 Data Terminal Connections

Procedure

Asynchronous Character Patterns

In this lab you will scope the signals coming out of the DTE.

Use the jumper cables that come with the breakout box (BOB) to connect the oscilloscope to pin 2 of the BOB. Connect the ground clip to pin 7 of the BOB.

Set the oscilloscope for positive edge trigger and DC coupling.

Make sure the **VAR** controls in the center of the vertical and horizontal controls are set to their **OFF** positions.

What voltage does the LED on pin 2 indicate for an idle condition? Does this correspond to the voltage that appears on the oscilloscope?

If your BOB is battery powered, turn the BOB off to save the battery. We do not need the LED indications for the rest of this lab.

Press the **CAPS** lock in the lower left corner of the keyboard so all the characters you type will be in upper case.

Use the same procedure as in the previous lab to set your data terminal or PC to have the following communication settings:

```
Data Bits   = 8
Bit 8       = 0
Parity      = OFF
Parity bit  = ODD
Stop bits   = 1
```

Press the letter C. The ASCII code for C is 67 Hex or 1 0 0 0 0 1 1 in binary. The 8th bit is a 0 according to the Bit 8 field previously shown. This gives us a bit pattern that looks like this:

0 1 0 0 0 0 1 1

The least significant bit gets transmitted first. This is the bit on the right of the pattern shown above. Preceding this bit is a start bit which is a 0 to distinguish it from the idle condition, which is a continuous string of 1's. Following the character string of 1's and 0's is the stop bit, which is always a 1.

The character pattern with its start and stop bits added appears below.

1 0 1 0 0 0 0 1 1 0

Keep in mind that the first bit sent down the line will be the start bit and the last will be the stop bit. This means the oscilloscope will trigger on the start bit and display it first on the left side of the scope display. The rest of the bits will follow in order so the pattern shown above actually is displayed in reverse order on the oscilloscope screen.

Draw the wave form for the letter C with its data format as shown above. Mark the voltage levels of the pulses on the drawing. Show two bits of idle condition before the start bit appears. Label all the bits.

1. What voltage is the:

   (a) MARK or binary 1? _____

   (b) SPACE or binary 0? _____

Calculate the frequency of the data bits by measuring the time between the 4 0's. Divide this time by 4 to get the time duration of a single bit.

2. What is the frequency of the bits? How does this compare to the value you set the terminal for?

Set the horizontal sweep rate to 0.5 msec. While pressing a character, adjust the **VAR** control in the center of the horizontal control so that 1 bit fills the distance between adjacent vertical lines. This will make it easier to count bits when you have a string of 1's or 0's.

Change the parity to ON.

With 8-bit characters, the parity bit will be the 9th bit.

3. What is the bit pattern for the letter C with 8-bit data characters and ODD parity? Show 2 idle bits and label all the other bits.

Draw your predicted wave form below.

Change the parity to even.

4. What is the bit pattern for the letter C with 8-bit data characters and EVEN parity?

Draw your predicted waveform. Show 2 idle bits and label the rest of the bits.

Change the character data bit to 7.

5. What is the bit pattern for the letter C with 7-bit data characters and EVEN parity? (Note: The parity bit becomes the 8th bit instead of the 9th bit when 7-bit data characters are used).

Draw the predicted waveform below for the situation in question 5. Show 2 idle bits and label the rest of the bits.

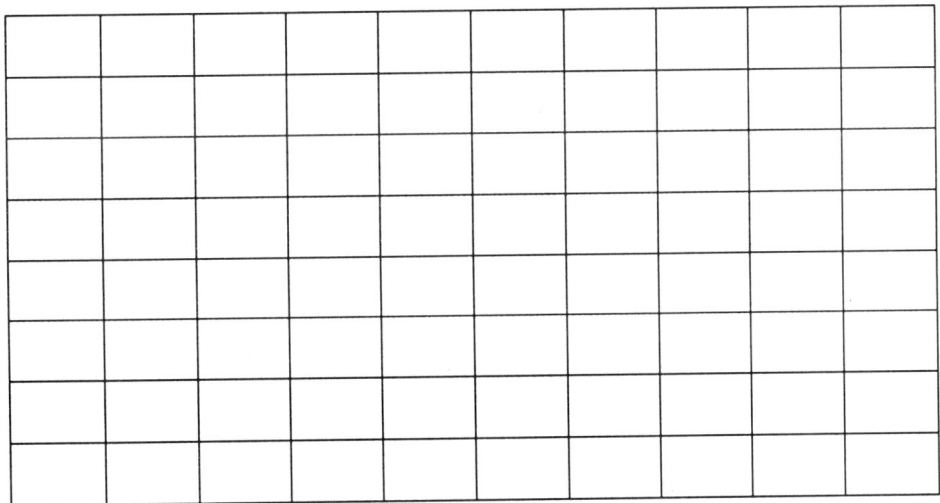

Use the oscilloscope to verify the results of the above questions.

RS-232 SIGNALS

LAB 19

Name: _____ Date: _____

OBJECTIVES:

Upon completion of this lab, you will be able to:

- Convert positive logic signals to RS-232 negative logic signals using a 1488 line driver
- Convert RS-232 negative logic signals to positive logic signals using a 1489 line receiver
- Measure and estimate the effect of line length on data transmission rates on an RS-232 data link

TEST EQUIPMENT:

- Dual-trace oscilloscope, function generator, DMM
- Low voltage dc power supply, low voltage dual dc power supply
- 1488 line driver, 1489 line receiver
- Capacitors (0.1 µF and 2500 pF)

Pre-Lab

Before you start making measurements, refresh your mind about some points of the RS-232 standard.

1. An RS-232 cable is connected to a DTE. On which pin would you find:

    (a) transmit data

    (b) receive data

    (c) signal ground

2. If the transmit data on an RS-232 cable were found to swing between +12 volts and −12 volts, identify at which voltage is the logic 1 and which voltage is the logic 0.

3. Over what range of voltages could a logic 1 be expected in the RS-232 standard? A logic 0?

Introduction

In answering the preceding questions, you should have noted that the RS-232 standard uses negative logic with a logic 1 being anywhere from –5 volts to –15 volts and logic 0 falling between +5 to +15 volts. The reason for this choice predates the development of modern electronics with its use of positive logic families such as TTL, LSTTL, and CMOS.

Inside the computer the logic families use positive logic with a logic 1 being close to +5 volts and a logic 0 being close to 0 volts. Often the electronics that an RS-232 signal must drive in the world outside the computer is also positive logic. This situation is illustrated in Figure 19-1.

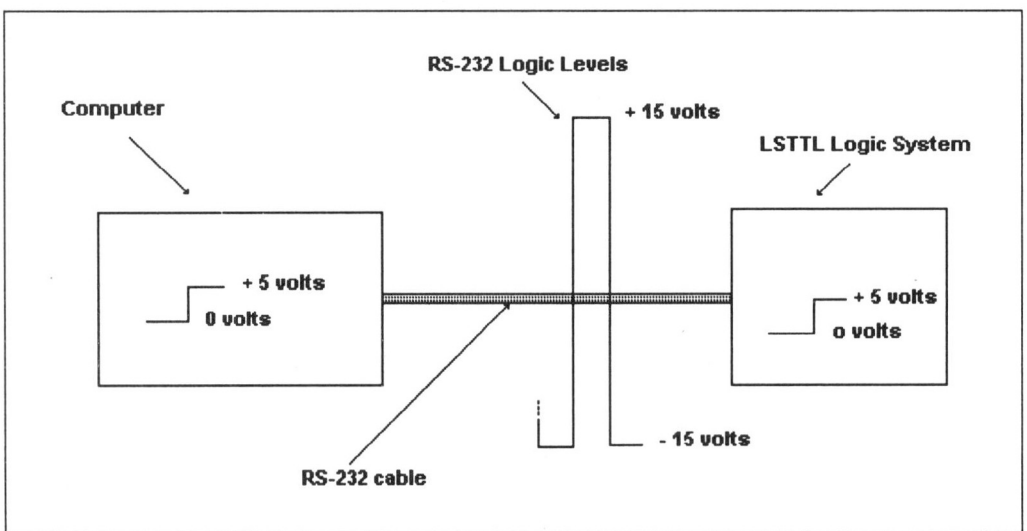

FIGURE 19-1

In order to deal with this problem, conversion circuitry is used. Two popular integrated circuits that were developed for this purpose are the 1488 line driver and the 1489 line receiver.

The 1488 line driver can be used to convert the positive logic levels of a LSTTL system to the negative logic levels of the RS-232 standard.

Connect the circuit shown in Figure 19-2.

4. Adjust the input voltage to the 1488 from 0 volts dc to +5 volts dc and monitor the output voltage.

You should have observed that positive logic to negative logic conversion took place.

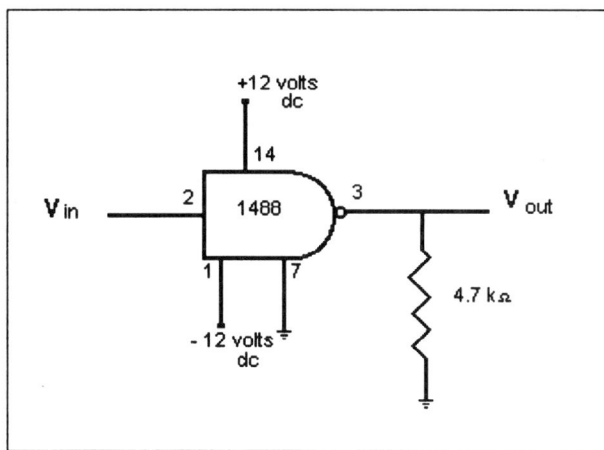

FIGURE 19-2

5. Comment on your observations. Was there conversion? What voltage did you measure for a logic 1 at the 1488 output? For a logic 0?

The 1489 line receiver, as the name indicates, converts the received RS-232 line signals to the desired level such as that of an LSTTL system. Connect the circuit as shown in Figure 19-3.

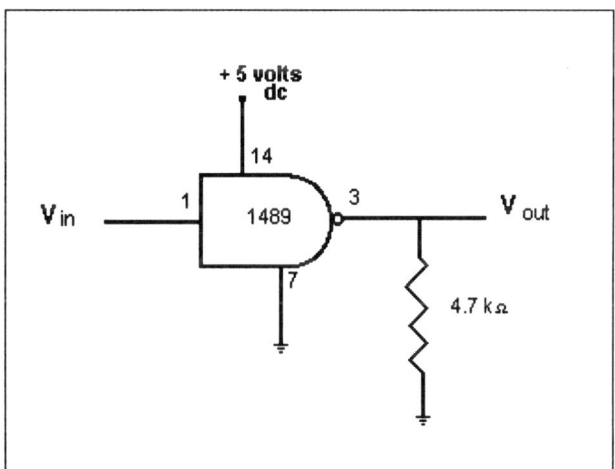

FIGURE 19-3

6. Adjust the input voltage to the 1489 from +12 volts dc to −12 volts dc and monitor the output voltage.

Again, you should have observed conversion—this time from negative logic to positive logic.

7. Comment on the results. Was there conversion?

Connect the output of the 1488 line driver to the input of the 1489 line receiver and use a 1000 Hz square wave swinging from 0 volts to +5 volts from a function generator to test out the circuit. Monitor the input to the 1488 with channel 1 of your oscilloscope and channel 2 to view successively the output of the 1488 (RS-232 levels) and the 1489 (LSTTL levels).

The circuit we have built replicates what happens to computer signals when they leave the electronics of a computer and travel down an RS-232 cable. The 1488 and 1489 integrated circuits are placed as shown in Figure 19-4.

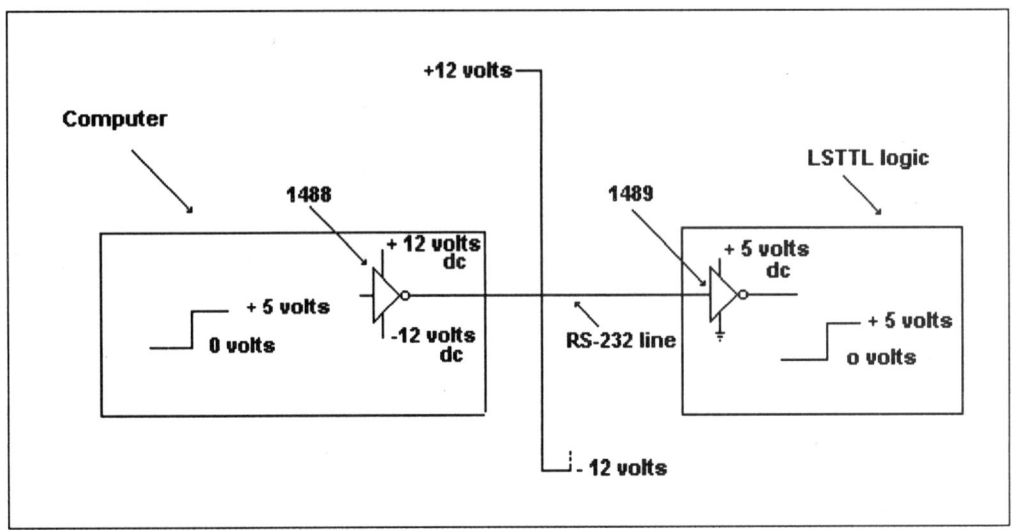

FIGURE 19-4

Before proceeding with more measurements, answer the following questions regarding the maximum data rate and maximum distance specified for data transmission on an RS-232 cable. Use your text or class notes as a reference.

8. What is the maximum specified data rate on an RS-232 cable?

9. Over what maximum distance does the standard recommend transmission at this data rate?

10. What role does the buildup of line capacitance play in setting the above data rate limits?

In researching the answer to these questions, you should have found that the longer the line the more capacitance there would be at the output of the 1488. As a consequence, the rise and fall times of the pulses would increase, causing data pulses to merge together. This limits data rates and line length.

Simulate the effect of having a long RS-232 cable by adding a 0.1 µF capacitor between the output of the 1488 line driver output and ground as shown in Figure 19-5.

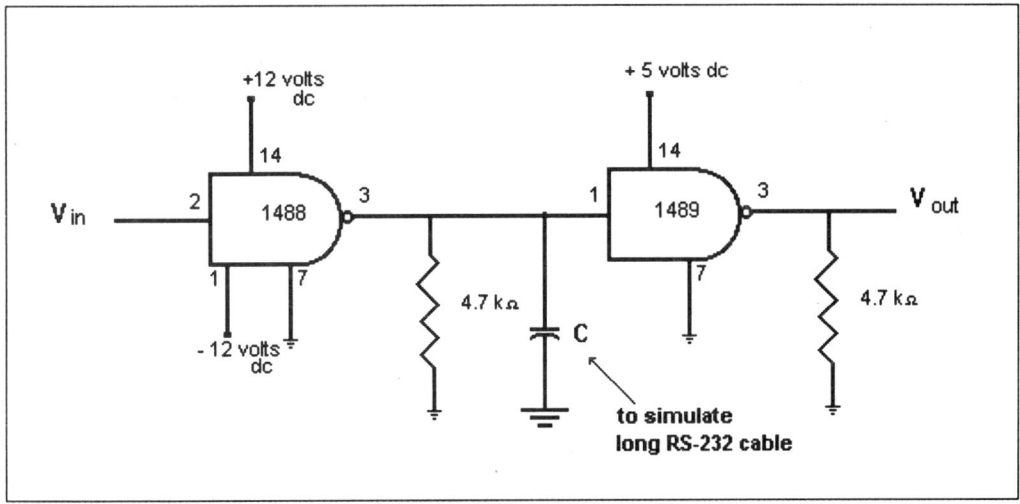

FIGURE 19-5

With the output of the square wave generator set to swing from 0 volts to +5 volts, vary the frequency of the square wave from 1000 Hz to 20,000 Hz and observe the output of the 1488 and the 1489.

11. What happens to the signal at the output of the 1489? Of the 1488?

12. As you increase the frequency of the generator is there a frequency at which the output of the 1489 does not change, indicating that the circuit is no longer working properly? What is this frequency?

Replace the 0.1 µF capacitor with a 2500 pF capacitor and repeat the above measurements. (2500 pF is the RS-232 standards' recommended maximum capacitance due to line length when operating at 20,000 bits per second.)

13. Comment on the results.

UARTS

LAB 20

Name: _____ Date: _____

OBJECTIVES:

Upon completion of this lab, you will be able to:

- Use diagnostic software to manipulate and monitor the registers of an Intel 8250 UART
- Interpret the status registers of an Intel 8250 UART
- Troubleshoot a PC serial communications system consisting of the internal serial port, the connector, the cable, and the device connected to the far end

TEST EQUIPMENT:

- PC diagnostic software, QAPLUS™ (version 4.52) installed in a directory named QAPLUS
- A computer capable of running QAPLUS™ and having at least one serial port using the Intel INS8250 UART
- RS-232 Tri-State Breakout Box and I/O Tester (DataCom Technologies Model 600 or equivalent)
- RS-232 25 pin cable
- Data sheets for the Intel INS8250 UART
- If possible, a serial I/O computer card with an INS8250 UART mounted on it in order to give a perspective on what a removable serial card looks like

Prerequisites

To fully understand this lab, the student should have received instruction on the following topics:

- Basic UART operation
- The Intel 8250 UART register set
- Highlights of the hardware interface signals out to the breakout box with special emphasis on **RTS, DTR, DSR, DCD, CTS, RI**

Procedure

QAPLUS™ is installed on the hard drive in a subdirectory called QAPLUS.

Figure 20-1 is a conceptual sketch of the lab equipment setup. Your instructor will provide you with a more detailed diagram specific to the particular equipment you are working with. With the **power to all equipment turned off** (which will remain off until your instructor has checked your setup and you have received a checkoff signature from your instructor), hook up the equipment as shown in the instructor's diagram. This diagram will show the equipment setup including where and how to

1. Connect the RS-232 cable to the RS-232 serial port at the back of the computer
2. Connect the other end of the RS-232 cable to the DTE side of the breakout box (BOB), making sure all the switches on the BOB are set to OPEN

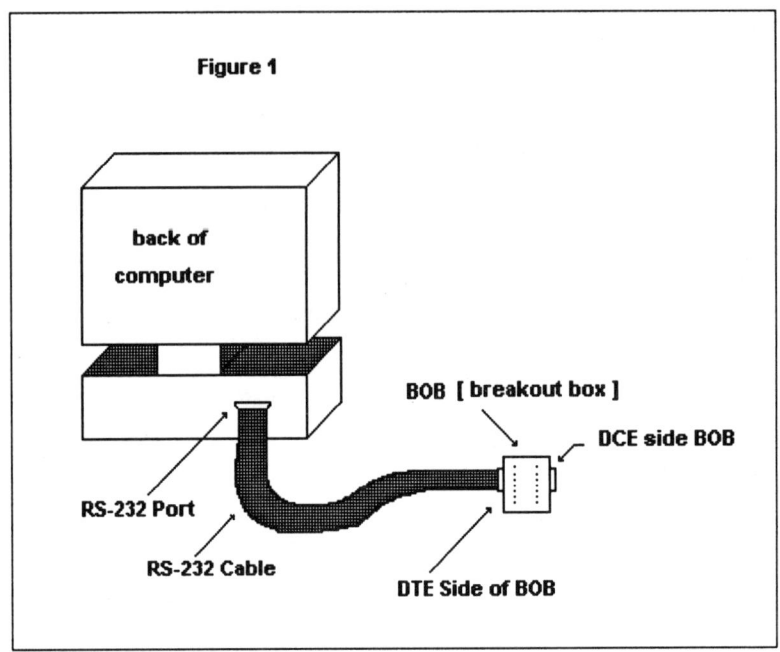

FIGURE 20-1

Have your instructor check your equipment setup.

Instructor's signature: _____

Now turn on the power to the equipment.

Turn the computer and monitor on. When the C>: prompt appears, type **QAPLUS\QAPLUS**. Use the left/right arrow key to highlight **Interact** on the menubar at the top. Use the up/down arrow key to select **COM debug**. Press the **Enter** key.

1. For each of the following hardware signals, explain what the letters stand for, what its function is, and indicate the direction of signal flow as either UART - > MODEM or MODEM - > UART. For the answers, refer to the INS8250 data sheets in Appendix B.

    (a) **CTS** stands for _____
    Function:

    Direction of signal flow _____

(b) DSR stands for _____

Function:

Direction of signal flow _____

(c) DCD stands for _____

Function:

Direction of signal flow _____

(d) RI stands for _____

Function:

Direction of signal flow _____

(e) DTR stands for _____

Function:

Direction of signal flow _____

(f) RTS stands for _____

Function:

Direction of signal flow _____

(g) BAUDOT stands for _____

Function:

Direction of signal flow _____

(h) SIN stands for _____

Function:

Direction of signal flow _____

(i) SOUT stands for _____

Function:

Direction of signal flow _____

UART Registers

The state of the hardware signals sent to the outside world is controlled by the **control register** settings of the UART. The software of the computer determines these settings.

The signals sent to the UART by the outside world are stored in the **status registers**. Use the INS8250 data sheets in Appendix B to answer the following questions.

2. How many registers does the INS8250 UART contain?

3. List the registers with a brief description of each register.

Now take time to read the function of the bits in each register. Do not worry if you do not understand every word, but a rough understanding will be useful for the next part of the lab.

4. What is the function of data bit 4 of the **modem control register?** See page 4–16 of the data sheets. Explain why you would want to set this bit. (Use your own words—don't copy what's in the data sheet.)

The software package that we are using allows us to investigate and modify the register contents of the UART in the PC.

5. What byte would you have sent (in HEX) to the **line control register** for 8-bit characters, with even parity, stick parity disabled, break disabled, and the system about to do a read operation of the divisor latches of the baud generator?

Diagnostic Testing

The screen in front of you should display the register set of the INS8250.

Set the COM port to COM2 by typing in COM2.

The 8250 can operate in local loopback mode if data bit 4 of the **modem control register** is set to a logic 1. Do this with the following command:

modemctrl = 10

In local loopback mode the UART will send data from its data-out register to its data-in register to allow you to test the equipment as far as the UART (without going to the outside world).

The following command will allow you to send 0F Hex to the receive register.

dataout = 0f

Did the same bit pattern appear at the data-in register? It should have if you've done everything correctly. Now try this command:

dump

6. What does the command **dump** do?

To get out of the local loopback mode we must change bit 4 of the **modemctrl register** to 0. When we are out of the local loopback mode, any data sent to the data-out register will be sent to the outside world.

7. What byte must be sent to the **modemctrl register** to get us out of the local loopback mode?

Type the command to get us out of the local loopback mode. The **modemctrl register** should have no bits set.

Clear the data windows by typing in **clear**.

Attempt the dataout command again. Type in:

dataout = 0f

Then try

dump

Note that what is sent out is not received. This is because there is not a complete loop. There is no device on the end of the RS-232 cable. Jumper the transmit data and receive data pins (2 and 3) on the BOB. Try both of the previous commands. This time the **datain** register should have received the data that was sent out.

When a DTE is ready to send data to a DCE, it must send a Request To Send (RTS) signal to the DCE.

8. What byte must be sent to the **modemctrl register** to activate the RTS line? Refer to the 8250 data sheet.

Type in the required command:

modemctrl = <answer to question 8>

If you entered the correct value, the RTS LED on the BOB should have gone red.

When the DCE is ready to accept the data from the DTE it sends the DTE a **clear-to-send** signal on the CTS line. This will cause the appropriate bit to be set in the **modem status register.** You can simulate the receiving of a CTS signal by placing a jumper from the RTS pin to the CTS pin. Do this.

9. What is the setting on the **modem status register?**

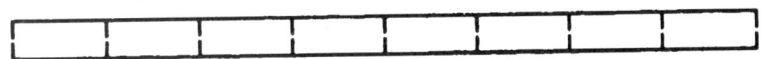

10. What bit does the 8250 data sheet say should be set to indicate the CTS line has gone active?

11. What is the command to set the DTR line active?

If you entered the correct command, the LED on the BOB opposite the DTR pin should have gone red. The modem would signal the DTE that it is powered up by making the Data Set Ready (DSR) line go active (red). Simulate that the modem is on by placing a jumper from the DTR pin to the DSR pin on the DTE side of the BOB.

12. What is the setting on the **modem status register?**

13. What bit does the 8250 data sheet say should be active to indicate that the modem is ready?

14. Simulate that the modem has detected a carrier by placing a jumper between the DTR pin and the DCD pin. What is the setting on the **modem status register**?

15. Simulate that the modem is being rung by placing a jumper between the DTR pin and the RI pin. What is the setting on the **modem status register**?

16. Explain how you could use the information you have just accumulated to diagnose a problem with a communications system. Is the problem the UART, the cable, or the peripheral device?

17. What is the current setting on the **line control register**?

18. From these settings, what do we know about:

   (a) Word length _____

   (b) Parity (odd/even/off) _____

   (c) Number of stop bits _____

19. What is the current setting on the **line status register**?

20. From these settings, what do we know about the status of the last received character as to:

   (a) Parity error _____

   (b) Overrun error _____

   (c) Framing error _____

21. What is the meaning of each of the 3 errors in question 20?

22. Examine the **Interrupt Enable Register**. Which bits are set?

23. What do the settings in question 22 tell us?

This concludes the lab. Exit the program by pressing the **Esc** key. When you have exited the program turn the computer and monitor off and remove the RS-232 cable.

MODEMS AND FILE TRANSFER PROTOCOLS

LAB 21

Name: _____   Date: _____

OBJECTIVES:

Upon completion of this lab, you will be able to:

- Use the Hayes command set to configure a modem
- Establish a communications link between two PCs using a communications software package, PROCOMM PLUS™, and a Hayes compatible modem
- Monitor and investigate a file transfer protocol in real time using a logic/data analyzer

TEST EQUIPMENT:

- Two computers with the capabilities of at least an IBM XT. They must each contain a serial port.
- Two modems capable of 1200 bps. They must be Hayes compatible.
- RS-232 Tri-State Breakout Box (DataCom Technologies Model 600 or equivalent)
- Three RS-232 25 pin cables
- One data analyzer (equivalent to TEK 308 logic analyzer)
- Communications software PROCOMM PLUS™ (version 2.01) for DOS

Prerequisites

To fully understand this lab, the student will have completed the labs that preceded this one and should have received instruction on the following topics:

- Modem communications and file transfer protocols
- Hayes command set for modems

Procedure

Figure 21-1 is a conceptual sketch of the lab equipment setup. Your instructor will provide you with a more detailed diagram specific to the particular equipment you are working with. With the **power to all equipment turned off** (which will remain off until you receive an equipment setup signature from your instructor), hook up the equipment as shown in the instructor's diagram. Remember you must plug the telephone line into the jacks marked "TO LINE." Also make sure that all pins dedicated for the RS-232 standard on the BOB are closed. If your BOB has another switch bank for other BOB uses such as automatic null modem, they must be left open. Your instructor will provide details specific to your breakout box indicating which switches are to be closed and which are to be left open.

Instructor's signature: _____

Now turn the power on to all the equipment. (Don't forget the modem.)

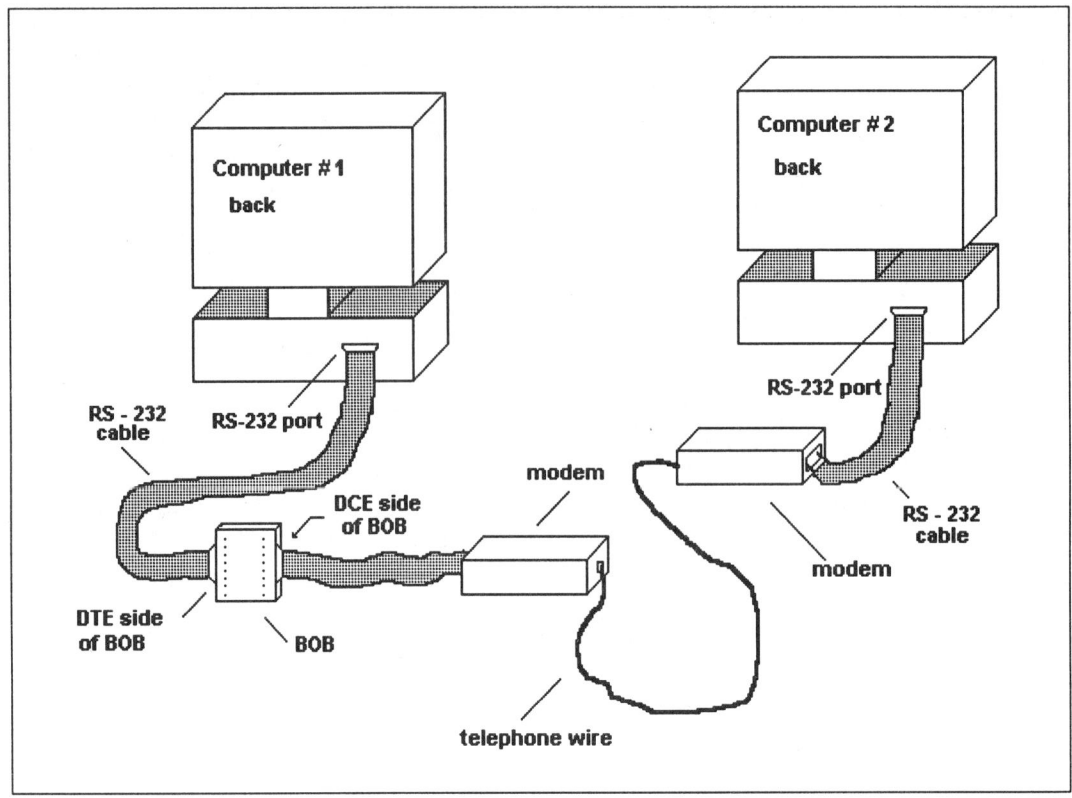

FIGURE 21-1

1. What LEDs are lit on the front of the modem? What are they telling you? Consult the modem manual.

Call up the communications software package PCPLUS™ from your computer.

Once PCPLUS™ is running, press any key to enter the terminal mode.

PROCOMM PLUS (PCPLUS)

PROCOMM PLUS is a communications package that allows a user to control the UART inside the computer. For one computer to talk to another computer, the two computers must be using the same protocol. In this lab we will be using the following communications parameters:

Bit rate	= 1200
Parity	= none
Bits/char	= 8
Stop bits	= 1

This is shown on the status line at the bottom of the computer screen.

LAB 21 MODEMS AND FILE TRANSFER PROTOCOLS 123

2. Which field contains the settings just discussed? What is in this field?

3. From the status line, are we operating in half duplex or full duplex mode?

The first field of the status line tells you to press **Alt-Z FOR HELP.** Hold down the **Alt** key and press the **Z** key at the same time. The menu that appears tells you how to access all the features available on PROCOMM PLUS.

We can change the settings of our communications parameters by calling up the **Line/Port Setup.**

4. What keys must you press to activate the **Line/Port Setup** menu?

Press the **Esc** key to return to the terminal screen. Press the key strokes required to call up the **Line/Port Setup.**

5. List the 5 different categories of line port parameters that you can change from this menu.

   (1)

   (2)

   (3)

   (4)

   (5)

Press the **Esc** key to return to the terminal screen.

In the UART lab preceding this lab, we used QAPLUS to change the bits of the registers in the UART. This allowed us to change the **Line/Port** settings.

LAB 21 MODEMS AND FILE TRANSFER PROTOCOLS

PROCOMM PLUS allows us to do the same thing but in a different manner. We can just select the settings from the last menu we looked at and the software changes the registers of the UART for us.

6. What LEDs have changed on the modem since question 1? Explain why the changes have taken place.

Turn the BOB on.

7. Explain what the colored LEDs on the DTE side of the BOB mean.

8. Open switch 20. What happens to the LEDs on the front of the modem? Explain why this has happened.

Close switch 20 on the BOB. Turn the BOB off to save the battery.

Type several words on the keyboard and press the enter key. Do it again a second time.

Notice that the new characters overwrite the last characters instead of appearing below the first line of characters you typed. With the current settings we generate only a carriage return when the **Return** key is pressed, instead of a carriage return and a line feed. Call up the HELP screen.

9. What key strokes must you press to toggle the **CR-CR/LF** so a carriage return and line feed are generated?

Press the **Esc** key to get out of the HELP menu. Toggle the **CR/LF** mode. Try typing a few lines of characters terminated with the **Return** key to ensure that each line of characters appears on a new line.

Press the **Return** key several times while watching the modem LEDs.

10. Which LEDs on the modem light up when you press the **Return** key? What do they mean?

Type out a message on one computer.

11. Did the other computer receive the message? Explain the results.

To communicate between two computers we must first make a connection between the computers via the modems. This is the same as you not being able to wish your Ma a Happy Easter on the phone until you dial her number. Just dialing Ma's number is not enough, as you may well know. There is some "handshaking" that must take place before you say, "Happy Easter, Ma."

After dialing the number, the phone must ring on the other end. Ma must pick up the phone and say "Hello." You must then say "Hello." This is a form of "handshaking."

This is basically what the modems must do to establish a connection. First ensure that the modem and computer are communicating properly with each other. To do this type **AT** and press the **Return** key. If the modem received the AT command (AT means attention), it sends back an **OK** to the computer.

The modem we are using has a small operating system built into a ROM chip, just as the computer has a set of instructions built into its ROM chip. We can send the modem commands and it will respond in a certain way. AT is one of these commands. The set of commands used by our modem is called the **Hayes** command set. Hayes commands for modems have become the de facto standard.

All the commands must be preceded by **AT** to alert the modem that a command is being sent to it. For instance, to dial the telephone number 256-9331 you would enter:

ATD2569331

Since we do not have enough telephone lines so everyone can use the modems over the telephone system, we will connect our modems directly to each other. Just imagine we are using the telephone system, since this is normally how you would use a modem.

We must still establish a connection between the two modems. There is a set of special Hayes commands that can be entered at the computers to allow us to establish a connection without dialing a phone number.

12. Refer to the modem manual. How do you establish a connection without dialing through the telephone system?

    Originate modem command = _____

    Answer modem command   = _____

    Establish a connection between the modems as described in question 12.

13. What was the noise you heard in the modems?

If you gave the proper Hayes commands, your computer should be displaying:

CONNECT 1200

Try typing on one of the computers. Does it show up on the originate-computer? Does it show up on the answer-computer?

14. How can you rectify the problem so you can see what you are typing on both computers at the same time?

Call up the HELP screen.

15. What keys must be pressed to implement the solution you proposed in question 14?

16. Now that you have established a connection between the two computers, what changes have taken place on the modem LEDs? Explain why the changes have occurred.

It is nice to be able to type messages back and forth to each other over the computers via the modems. This is handy if you are deaf. Otherwise, it would be easier and cheaper to just talk over the phone to the other party.

The true strength of this setup lies in its ability to transfer data from computer to computer. To transfer data, we require a file transfer protocol.

Xmodem File Transfer Protocol

The main reason for using a file transfer protocol is to ensure that data is received correctly. In all file transfer protocols there is some sort of checksum or CRC character(s) added to the end of the block of data that is sent. This special character(s) allows the receiving station to verify if the data was received correctly. If there was as much as a single bit of data changed while the packet was traveling from the transmitter to the receiver, the checksum or CRC will not work out. The receiving station then sends the transmitting station a message requesting that the last packet be sent again.

Xmodem is one of the most common file transfer protocols. Xmodem takes a block of characters (128 bytes or characters) and packets it up as shown below:

SOH	PACK #	COMP. of PACK #	DATA = 128 bytes	CRC 16

SOH	Means **Start of Header**. This identifies the beginning of a block.
PACK #	The blocks are numbered. This field contains the number of the block.

COMP. of PACK #	This is the complement of the block number. The receiver complements this number and compares it to the block number to make sure the number was received correctly.
DATA	This is the 128-byte message that was sent.
CRC 16	Xmodem uses CRC 16 for error checking. Other protocols may use other error checking algorithms.

PROCOMM PLUS has the ability to packet up the data in one of several different protocols. Call up the HELP screen.

17. What are the key strokes to:

   (a) Send a file _____

   (b) Receive a file _____

   Activate the **Send File** option on one of the computers.

18. The most common file transfer protocols are listed as **X), Y), Z) and K)** on the **Send File** screen. What are the full names listed opposite these letters?

   X) = _____

   Y) = _____

   Z) = _____

   K) = _____

   Select **X** for Xmodem. Enter **AUTOEXEC.BAT** when you are asked to **Please enter filename:**. (You can use upper case or lower case). Make sure you press the **Return** key after entering the filename.

19. How large is the file in bytes?

20. How many blocks will be sent?

21. Why does it take this many blocks to send the file?

   On the other computer, activate the **Receive File** option. Select the Xmodem again. For **filename**, enter **TEST.BAT**.

➤ *NOTE:* The transfer may be aborted while you are answering the questions. If you do not complete the send/receive operations within a certain time limit, the operation is aborted. That is not a problem; you can just initiate the operation again.

22. What is the value opposite **block count** on the receiving station?

23. Why are the number of bytes in the file (question 19) and the number of bytes received (question 22) *not* the same?

You can view the directory of the receiver station by activating the **File Directory** option. This will prove that AUTOEXEC.BAT was transferred from one computer to the other.

24. Use the HELP screen to determine which keys must be pressed to activate the **File Directory** option. What are the keys?

Activate the **File Directory** option on both computers. Did the file get copied to the receiver station? (You may have TEST.BAT on both computers from an earlier running of the lab, but the date and time opposite the filename will help distinguish which file has just been copied over the system.)

Notice that AUTOEXEC.BAT has one size shown and TEST.BAT has another.

We will now use the data analyzer to capture the blocks of data that are sent from the transmitting computer to the receiving computer via Xmodem. The rest of the lab assumes the transmitting computer is connected to the BOB while the receiving computer is connected directly to the modem.

25. What pins must the data analyzer be connected to on the BOB to capture the transmitter's packages?

Your instructor will provide you with instruction sheets for the data analyzer that you are using.

26. Show the Hex, binary, and ASCII character representations of the first 3 bytes of the Xmodem package.

If your data analyzer puts < > around "E" in <E> on the screen of the data analyzer, this means the letter "e" was sent, not the letter "E." The data analyzer will do this if it only displays upper case letters.

27. Refer back to the fields of an Xmodem package. What are the first 3 characters of the package? Do the first 3 bytes of the captured message correspond in any way to the first 3 fields of the Xmodem package?

You should be able to see recognizable commands on the logic analyzer which correspond to AUTOEXEC.BAT instructions.

Notice you can even see the carriage returns (CR) and the line feeds (LF) that were generated whenever the Return key was pressed at the end of each line of AUTOEXEC.BAT.

Continue scrolling through the memory buffer of the data analyzer. The second block begins with **SOH**.

28. What are the two bytes just before SOH? What do they represent? (Refer to the fields of the Xmodem package).

29. What are the next two bytes following SOH? Are they what you expected?

Xmodem is an **Automatic Repeat Request (ARQ)** protocol. This means the transmitter sends a package and the receiver checks the package to make sure it was received correctly. If the package was received correctly the receiver sends the transmitter a positive acknowledgement message. If an error was detected in the package the receiver sends a negative acknowledgement. Xmodem uses the character **ACK** as a positive acknowledgement and the character **NAK** as a negative acknowledgement.

30. What pins on the DTE side of the BOB must the data analyzer be on to capture the acknowledgement signals from the receiving computer?

Connect the data analyzer to the pins on the BOB that you listed in question 30. Start the data acquisition process on the data analyzer. Send AUTOEXEC.BAT again, using Xmodem.

Now we will be using the data/logic analyzer to capture the control characters that the *receiving* computer sends at the start of and during the file transfer process with XMODEM as the file transfer protocol.

Since we will not be capturing a full buffer of data as we did before when we used the logic analyzer, the procedure for using the analyzer will be slightly different. The instructions will vary depending on your analyzer. In any case, your instructor will provide them for you as before.

If everything went as planned you should have captured:

ACK

ACK

NAK

ACK

The first two ACKs are positive acknowledgement of having received the two packages. When all the packages were sent, Xmodem sent two EOT (End of Text) characters to signal the end of transmission. The receiving station sent a NAK in response to the first EOT because it did not recognize it as anything useful or correct. The second EOT is the Xmodem signal that the end of transmission has been reached. The receiving station sent an ACK to acknowledge it.

Perhaps you could not see the EOT characters while monitoring the transmitter signal because the buffer on your logic analyzer is too small and filled up before you encountered them.

You could capture the characters if you set up the analyzer to ignore all characters before a special two character string. Consult the manual for the analyzer on how to do this, or perhaps your instructor will provide these for you. You could then enter the two characters for package number 2 and its complement. This would allow you to capture the complete second package and the two EOT characters.

Move the data analyzer probe to pin 3 on the DCE side of the BOB. Open the switch for pin 3. Now when we send a package using Xmodem, the receiver will send an **ACK** on pin 3, but the transmitter will not get the ACK because the line is open. When the Xmodem transfer "times out," view the captured data.

31. What signals did the receiver send to the transmitter while pin 3 was open on the BOB?

This concludes the lab. Exit out of PCPLUS. Switch pin 3 to the closed position on the BOB and make sure the battery switch on the BOB is OFF.

PULSE CODE MODULATION AND CODECS

LAB 22

Name: _____ Date: _____

OBJECTIVES:

Upon completion of this lab, you will be able to:

- Measure signals in the transmit and receive section of a PCM link
- Transmit and receive a signal over a PCM link

TEST EQUIPMENT:

- Pulse code modulation trainer board by *Lab-Volt®**
- Dual-trace oscilloscope
- Low voltage dc power supply capable of ±15 V dc
- Audio function generator, Interstate F43 or comparable

*PCM trainer board (model # AS19481) available from *Lab-Volt®* Ltd. (1-800-522-8658)

Purpose

This lab is designed to help you investigate how pulse code modulation (PCM) operates. You will record signals at various points in the receiver and transmitter sections of a codec.

Introduction

Pulse code modulation is the process of converting an analog signal into a serial stream of 1's and 0's. The coded binary signal represents the amplitude of the analog signal. The process also involves converting the serial binary signal back into an analog signal.

The electronic device that is most commonly employed for this purpose is the codec.

Codec stands for coder/decoder.

This device contains a

1. Transmitter that converts the analog signal into a coded binary signal
2. Receiver that converts the coded binary signal back into an analog signal

The codec used in this lab is the TP3020. It consists of a

1. Transmitter with
    a. Sample-and-hold circuit
    b. ADC (analog-to-digital converter)
    c. Parallel-to-serial encoder

2. Receiver with
   a. Regeneration circuit
   b. Serial-to-parallel decoder
   c. DAC (digital-to-analog converter)
3. Timing and control circuitry to synchronize the encoder and decoder

PCM Filters

The TP3020 codec is matched to a transmitter/receiver filter chip called a TP3040. The transmitter filter is a bandpass filter designed to remove low frequency hum and high frequency signals above the voice range.

The receiver filter is a low pass filter that shapes the PAM code from the DAC into an analog signal.

1. Locate the TP3020 codec and the TP3040 filter ICs on the PCM trainer board. There are two sets of each chip.

   Where on the board is each chip?

   Filter for PCM CODEC 1

   Filter for PCM CODEC 2

   Codec for PCM CODEC 1

   Codec for PCM CODEC 2

2. How many pins does the TP3040 have?

3. How many pins does the TP3020 have?

Function of the Transmitter Blocks of the PCM System

Transmitter Filter

The audio signal is fed into the transmitter filter section of the TP3040 IC. This bandpass filter allows only the signals that fall within the audio range to reach the codec.

Sample-and-Hold

The analog-to-digital conversion process takes a finite amount of time. If the ADC tried to convert the input signal "on the fly," the binary codes would be invalid. The input signal must be sampled and latched at the sampled level. The ADC converts the latched-sample to a parallel digital-binary code.

For instance, a sample of the input signal may be taken when the input signal is at 2.35 volts. The sample-and-hold circuit latches at 2.35 volts. While the ADC converts the 2.35 V latched-signal to a digital code, the input signal may have changed to 2.44 volts.

The output of the sample-and-hold circuit is a pulse amplitude signal, a PAM.

ADC (Analog-to-Digital Converter)

The ADC converts the PAM code from the sample-and-hold circuit to a series of parallel bits.

Encoder

This part of the codec converts the parallel output of the ADC into a serial stream of bits that can be transmitted over a two-wire medium.

4. Draw the block diagram of the transmitter section of the codec with the transmit filter connected to it. Label each of the 4 blocks (including the filter as one of the 4 blocks) and draw the waveform at each output.

Functions of the Receiver Blocks of the PCM System

Regeneration Circuit

This circuit reshapes the serial input so the next stage has "clean" bits to work with.

Decoder

This part of the receiver converts the serial binary input into a parallel input for the DAC.

DAC

The DAC converts the parallel binary input into a PAM code that roughly approximates the original analog signal.

Receiver Filter

This is a low pass filter that converts the PAM code from the DAC into a smooth analog signal.

5. Draw the block diagram of the receiver section of the codec with the receive filter connected to it. Label each of the 4 blocks (including the filter as one of the blocks) and draw the waveform at each output.

Frequency Range of the Filters

In this part of the lab you will inject an audio signal into the transmitter filter section of the TP3040 IC-filter. You will then measure the output for different input frequencies and sketch the gain vs. frequency curve for the filter.

You will then repeat the procedure for the receiver filter section of the TP3040.

6. Draw the equipment setup for this part of the lab. Refer to the explanation of how to set up the equipment that follows this question.

Equipment Setup for Measuring the Transmitter Filter

Connect ±15 V dc to the PCM trainer (at the top of the board). **Do not turn the power supply on yet.**

Connect J21 of the trainer to J1 of the trainer. This will supply the timing signals required by the filter.

Connect the audio generator output to J13 of the PCM CODEC 1 section of the PCM trainer board. This is the input to the transmitter filter.

Connect channel 1 of the oscilloscope to J13. Set the oscilloscope for 0.2 V/cm with a X10 probe connected to it. Set the sweep speed for 0.5 ms/cm. Make sure both channels are on "DC" input.

Turn the power supply on.

Set the audio generator for 1 kHz and 4 V p-p on the oscilloscope.

Make sure the verniers marked "var" on the vertical and horizontal controls are set to their maximum clockwise position, otherwise your readings will be inaccurate.

7. What is amplitude of the signal at the output of the transmit filter (J14)?

   _____ V p-p

8. Calculate what the amplitude of the signal will be at the 3 dB-down point?

   _____ V p-p

   While measuring the output of the transmit filter at J14, adjust the frequency of the audio generator until the output falls off to the −3 dB point that you calculated in question 8.

   Record the low frequency 3 dB point: _____ Hz

   Next, increase the frequency of the audio generator until the output falls off to the −3 dB point calculated in question 8.

   Record the high frequency 3 dB point: _____ Hz

9. Sketch the gain vs. frequency curve for the transmit filter. Indicate the frequencies at the 3 dB points.

10. Why does the transmit filter have the frequency characteristics shown in question 9?

Move the audio generator to J19 of the PCM CODEC 1. This is the input of the receiver filter.

Connect channel 2 of the oscilloscope to J20. This is the output of the receiver filter.

Adjust the audio generator for 6 V p-p at a frequency of 1 kHz.

11. What is amplitude of the signal at the output of the receive filter?

    _____ V p-p

12. Calculate what the amplitude of the signal will be at the 3 dB-down point relative to the passband signal measured in question 11.

    _____ V p-p

    While measuring the output of the receive filter at J20, adjust the frequency of the audio generator until the output falls off to the −3 dB point that you calculated in question 12.

    Record the low frequency 3 dB point: _____ Hz

    Next, increase the frequency of the audio generator until the output falls off to the −3 dB point calculated in question 12.

    Record the high frequency 3 dB point: _____ Hz

13. Sketch the gain vs. frequency curve for the receive filter. Indicate the frequencies at the 3 dB points.

14. Why does the receive filter have the frequency characteristics shown in question 13?

Signal Measurements on the Codec

In this part of the lab you will inject a signal into the transmitter section of the CODEC 1. The output of CODEC 1 will be connected to the receiver of CODEC 2. You will measure the signal as it gets converted from an analog signal to a digital signal and then back into an analog signal. The diagram in Figure 22-1 gives you the conceptual idea, while the diagram in Figure 22-2 will show you what connections to make.

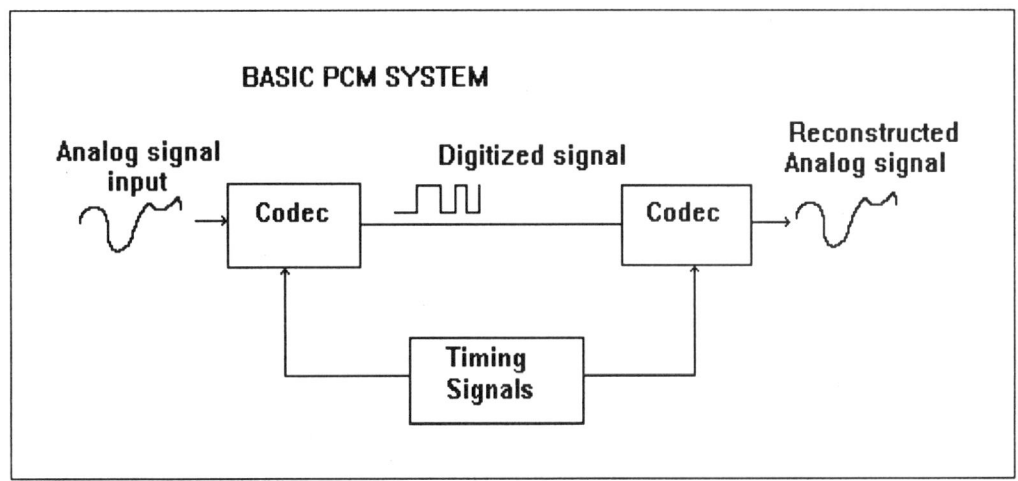

FIGURE 22-1 Basic PCM System

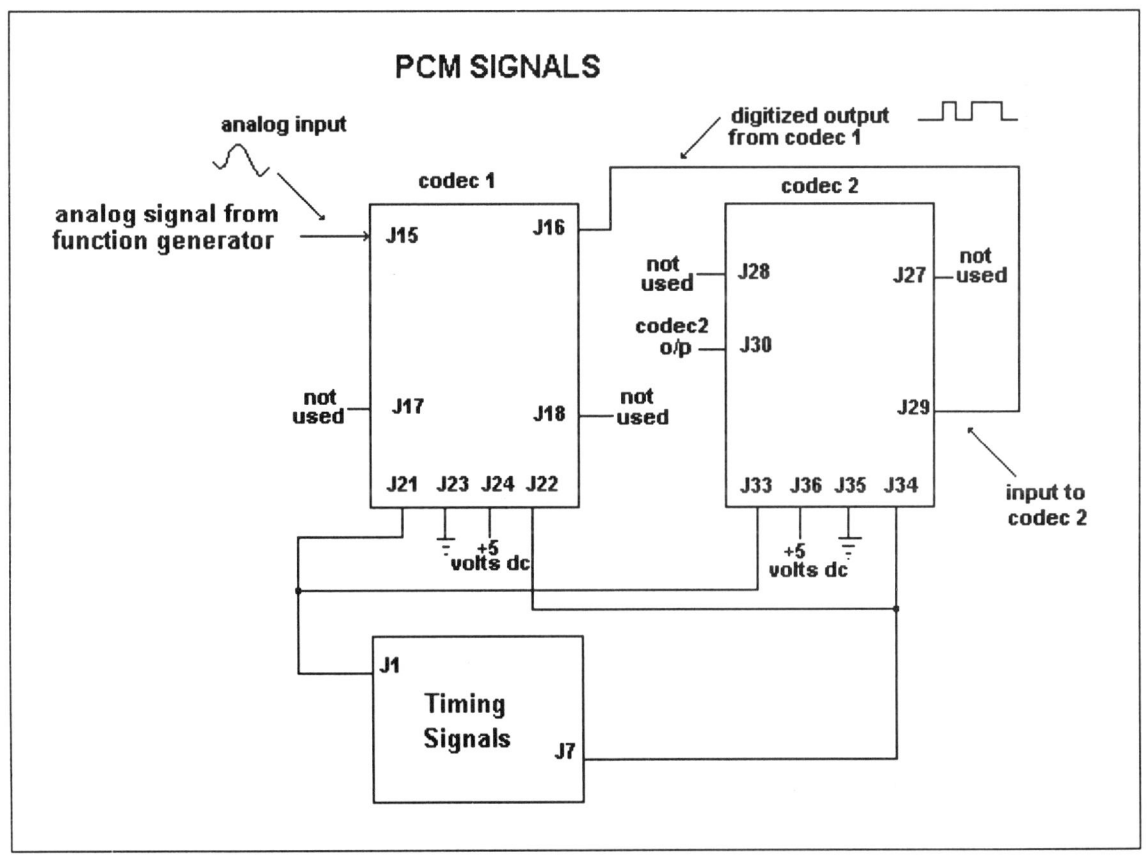

FIGURE 22-2 Setup for Measuring the Signals in a PCM System

Connect the equipment as shown in Figure 22-2.

Notice pins J24 and J36 must be connected to +5 V dc. The +5 V dc socket is located on the PCM DECODER section of the PCM trainer. The PCM DECODER section is located in the upper right corner of the trainer.

Set the audio generator up for 4 V p-p at 1 kHz. Connect channel 1 of the oscilloscope to J15. J15 is the input to CODEC 1. The output of CODEC 1, J16, is connected to the input of CODEC 2, J29. The output of CODEC 2 is J30. **Label the J's in Figure 22-1.**

Turn the power supply on.

Connect channel 2 on the oscilloscope to J16, the output of CODEC 1.

Turn off channel 1 and set the SOURCE trigger to CH 2.

Set the horizontal sweep for 50 usecs/cm.

Set the vertical gain control to 0.2 V/cm.

15. Draw the waveform at J16. Indicate the amplitude and period. Calculate the frequency.

period = _____

freq = _____

Try adjusting the frequency of the audio generator while recording any changes in the output of CODEC 1. Try adjusting the amplitude of the input signal from the audio generator while recording the changes at the output of CODEC 1.

16. What effect on the output of CODEC 1 occurred from

   (a) changing the amplitude of the input signal?

   (b) changing the frequency of the input signal?

It seems strange that the output of the codec is not affected by the change in frequency or amplitude of the input signal that is being digitized.

Actually the signal is being digitized. Each pulse you drew in question 15 is made up of 8 bits representing one of the samples of the input signal.

Turn the horizontal sweep to .5 usec/cm. You will not be able to stabilize the display since the 7-bit pattern that is displayed is constantly changing as the input signal fluctuates by small amounts. The ADC is actually an 8-bit device but the first bit being a 0 for negative voltages does not show up since the oscilloscope triggers on the first positive bit.

17. Draw the waveform at J16 that represents the bits being transmitted from CODEC 1.

Getting back to question 15, the waveform was a series of pulses, each consisting of 8 bits. If you measured the pulses correctly, you should have measured a frequency of 8 kHz in question 15.

18. Since each pulse of the 8 kHz signal consists of an 8-bit binary coded signal how many bits are transmitted per second from the codec?

19. What is the clock frequency provided by the TIMING SIGNAL GENERATOR section of the PCM trainer? (You can find the clock frequency on the crystal.)

A bit can be generated for every one of the 1,536,000 clock pulses. The first 8 clock pulses result in an 8-bit sample being transmitted. Then there is a long delay before the next 8-bit sample is transmitted. If there was another codec transmitting its signals, it could be multiplexed on the same line by timing it so its 8-bit sample would fall in the gap between the 8-bit samples of CODEC 1. This is the standard value used in telephony.

The clock frequency is 1.536 MHz and each channel requires 8 x 8 kHz bits. Therefore, it is possible to multiplex 1.536 MHz/64 kHz = 24 channels.

Move the channel 2 probe to the output of CODEC 2, J30. Set the horizontal sweep for 0.2 msec/cm. Set the vertical gain for 1 V/cm. Make sure the audio generator is set to 1 kHz at 4 V p-p.

You will not be able to perfectly stabilize the signal. Use the LEVEL control in the upper right corner to give the most stable pattern possible. Try making small adjustments of the frequency of the audio generator to help stabilize the pattern.

20. Draw the waveform at J30, the output of CODEC 2.

21. How many samples are there per cycle of the input signal?

22. Explain why the number of samples in question 21 is what you measured.

Try adjusting the amplitude of the input signal. You should be able to see the digitized signal change. Try changing the frequency of the input waveform. Test the validity of the sampling theorem by increasing the input waveform frequency above 4000 Hz.

23. What do you observe? Why?

Overload Distortion

Overload distortion occurs when the input signal exceeds the maximum allowed input of the encoder. Place the probe for channel 2 of the oscilloscope on the output of CODEC 2, J30.

Increase the output of the signal generator by changing the output control to 0 dB. Adjust the variable gain control up and down while observing the waveform at J30.

24. How can you tell when overload distortion takes place? Draw the PAM-code waveform showing overload distortion.

25. What is the maximum p-p voltage of the input signal that *does not* produce overload distortion?

When the input signal drops below the minimum detectable level of the encoder the output of CODEC 2 will drop to 0. Measure the minimum detectable signal by turning the output of the signal generator all the way down. While channel 2 is monitoring the output of CODEC 2 and channel 1 is monitoring the input signal to CODEC 1, increase the output of the audio generator until you just begin to notice the step-pattern characteristic of the PAM code on channel 2.

26. What is the minimum detectable signal?

_____ V p-p

Effect of the Receiver Filter

Earlier in the lab we measured the bandpass of the TP3040 filter chip. We did not connect the filter to our codec yet.

There is no need to connect the transmitter filter since its purpose is to eliminate signals outside the voice bandpass. We are using a clean signal from the audio signal generator that falls within the voice bandpass.

The low pass receiver filter has quite a marked effect on our received signal. We will take a look at that now. We have been measuring the PAM code from the output of CODEC 2 on J30. Let's run that signal through the low pass receiver filter. Connect J30 to the input of the receiver filter, J31. Connect channel 2 of the oscilloscope to the input of the receiver filter, J31, and channel 1 of the oscilloscope to the output of the receiver filter, J32.

27. Draw the signals at the input and output of the receiver filter.

You should have seen a reconstructed waveform that matches closely the input signal waveform.

DIRECTIONAL COUPLERS AND VSWR

LAB 23

Name: _____ Date: _____

OBJECTIVES:

Upon completion of this lab, you will be able to:
- Measure the return loss due to a mismatch
- Calculate the VSWR due to a mismatch

TEST EQUIPMENT:

- Two 50 ohm termination loads, 50 ohm coaxial cable
- Spectrum analyzer with tracking generator, directional coupler
- Optional: Video display monitor for demonstration

Prerequisites

To perform this lab, the student should have had instruction or prior exposure to the following topics:

- Directional Couplers
- VSWR and Return Loss
- Basic measurements with the spectrum analyzer

Introduction

Directional Coupler Concept

A directional coupler is a passive device which will couple power flowing in one direction to an auxiliary output. Power flowing in the opposite direction through the coupler will not have (ideally) any of its power coupled to the auxiliary output.

For example, suppose that 10 dBm (10 mW) of power were flowing into the IN port and out of the OUT port of the directional coupler shown in Figure 23-1. In other words, 10 dBm of power is flowing through the directional coupler in such a direction so as to couple some of this power to the auxiliary output, which in this diagram is labeled CPL. The degree of coupling is measured in dB and is called the Coupling Factor. If the coupling factor were 20 dB, the amount of power that would appear at the CPL output would be –10 dBm or 0.01 mW. This is arrived at by the following calculation:

Power at CPL or auxiliary output = Power flowing in dBm – Coupling Factor in dB

(+10 dBm) – (20 dB) = (–10 dBm)

If power were to flow through the directional coupler in the opposite direction (from OUT to IN), no power would be coupled to the CPL port.

VSWR Concept

The Voltage Standing Wave Ratio (VSWR) is a measure of how well matched a transmission line is to its load. For example, if an antenna of 50 ohms were connected at the end of a 50 ohm transmission line, there would be no reflected power and no standing waves.

Measurement of VSWR Using a Spectrum Analyzer and a Directional Coupler

The equipment hookup in Figure 23-1 illustrates how the VSWR of a load might be observed over a frequency range.

The tracking generator in Figure 23-1 is a signal generator whose frequency is constantly varying over the span of frequencies of the spectrum analyzer and is synchronized with the spectrum analyzer display.

Notice that the directional coupler is connected so that the tracking generator output is connected to the directional coupler port marked OUT and the device whose VSWR we want to measure is connected to the directional coupler port marked IN.

The directional coupler is designed so that signal power sent from the tracking generator into the OUT port goes through the coupler to the IN port and out to the device whose VSWR we want to measure. If the device (antenna, etc.) has a poor VSWR, there will be significant reflected power which will have no choice but to re-enter the directional coupler, this time through the IN port. As was noted before, not all the power entering a directional coupler via the IN port goes to the OUT port. A small fraction, say 1% of this power, is coupled to the port marked CPL and the remaining 99% to the OUT port.

When the device being tested for VSWR is a good match, there is little or no reflected power so very little power comes out of the CPL port. When the device being tested reflects a significant amount of power and so has a poor VSWR, the power from the CPL port increases.

Note that the port marked CPL is connected to the spectrum analyzer input.

Monitoring this power from the CPL port on the spectrum analyzer allows us to see at which frequencies we have a good match.

A measure of how well a mismatch reflects power is the Return Loss.

When all the incident power at a mismatch is reflected, the return loss is 0 dB.

For example, if there is no load connected for test, there is an open circuit and 100% of the tracking generator power is reflected to the IN port. The output from the CPL port in such a case would be proportional to a 0 dB return loss.

Procedure

Connect the equipment as shown in Figure 23-1, but do not connect a load to the IN port. Use 50 ohm coaxial cable for connections. **No power should be on at this time.** Have the instructor check your setup before turning on any power.

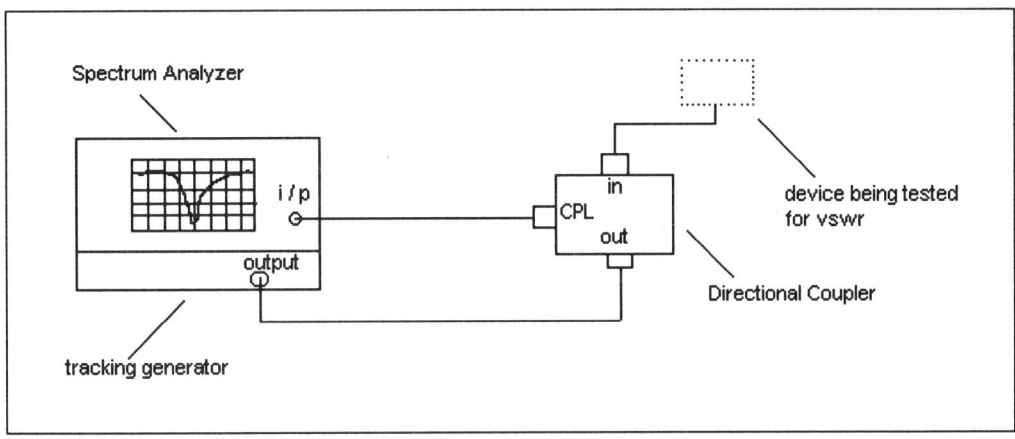

FIGURE 23-1

Instructor's signature: _____

Now turn the power on to all the equipment.

Set the desired center frequency and span of the spectrum analyzer.

At the moment you do not have a load connected to the coupler, so all of the power is reflected from the load under test. The analyzer should display a line near the top of the screen.

Adjust the vertical position control so that the screen trace intersects with one of the analyzer screen's upper horizontal graticules. We will call this the calibration trace and it sets the level on the screen for an open circuit or a 0 dB return loss.

Now connect one of the 50 ohm loads to the IN port. Presuming a good match, there will be little reflected power and the screen trace on the spectrum analyzer should drop, indicating a low VSWR over the frequency range set by the span of the spectrum analyzer. A VSWR of 1.0 (perfect match) would give a trace at the bottom of the screen and be buried in the "grass" or the noise of the analyzer.

1. Sketch your calibration trace and 50 ohm match trace. (Also indicate analyzer settings on this sketch.)

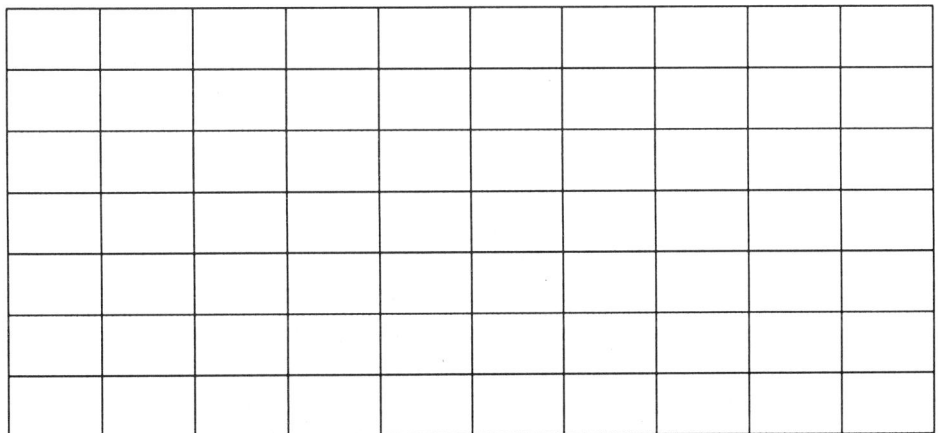

The difference between the calibration trace (0 dB line) and the trace caused by the 50 ohm load is measured in dB and is the return loss of the 50 ohm load being tested.

2. What was the measured return loss for the 50 ohm load?

Transmission line theory gives us a formula for calculating VSWR when we know the return loss in dB.

$$VSWR = \frac{10^{\frac{RL}{20}} + 1}{10^{\frac{RL}{20}} - 1}$$

where

RL = Return Loss in dB

3. Calculate the VSWR of the 50 ohm load.

You should have calculated a VSWR fairly close to the ideal value of 1.0. Replace the 50 ohm termination with two 50 ohm terminations in parallel, creating a 25 ohm load and a known mismatch.

4. What is the measured VSWR for the 25 ohm load?

5. What VSWR would theory predict? Show your work.

Often a mismatch is at a remote location such as a rooftop or at the top of a tower. The measurement of the VSWR due to the mismatch in these circumstances takes place at the end of a lossy coaxial cable far from the actual mismatch. The measured results in such a situation would indicate a much better situation than actually exists at the end of the coaxial cable.

Measure the VSWR of a 25 ohm mismatch when the 25 ohm load is at the end of a 50 foot length of RG-58A/U. Use the same frequency range as in 4.

6. What VSWR did you measure?

7. How did it compare with the VSWR measured in 4?

Answer: **You should have observed that the 50 feet of lossy cable masks the true amount of mismatch that exists at the end of the line.**

8. Look at a 50 Ω load at the end of one meter of 75 Ω cable. Explain what you observe.

TIME DOMAIN REFLECTOMETRY (TDR) — LAB 24

Name: _____   Date: _____

OBJECTIVES:

Upon completion of this lab, you will be able to:

- Determine the length of a coaxial cable from TDR measurements
- Determine the effects of terminating a transmission line with loads other than the proper termination

TEST EQUIPMENT:

- Pulse generator (TEK 114 or equivalent)
- Dual-trace oscilloscope
- Coaxial cable   as available (RG58, RG59, etc.)
                  3 meter and 25 meter length per station

Introduction

In this lab we will investigate the use of time domain reflectometry (TDR) in testing of cables. When a pulse is sent down a transmission line which is terminated with an impedance different from the characteric impedance of the line, a reflected pulse is set up at the impedance mismatch. The essential idea in TDR is that, by sending a pulse down a transmission line, we can measure on an oscilloscope the time from when the pulse was sent to the time when the reflection returns. If we know the speed of a pulse on the line, it should be possible to calculate the length of the line. Often transmission lines are broken by industrial activity. By the use of TDR we can locate exactly where the break is. For example, if the cable were behind a wall or buried under the earth, much time is saved in precisely locating the fault. Many companies provide specialized equipment for time domain reflectometry, but we can do just as good a job with a scope and a pulse generator.

Procedure

Before we proceed with measurements, we need some data for the cable we will be using. Your instructor will provide you with the specification sheets for the coaxial cable you will be testing. From the specification sheets, find or calculate the following information:

1. What is the nominal speed of propagation?

151

2. What is the characteristic impedance of the cable?

3. What is the nominal attenuation in dB/100 ft.? (Assume a frequency of 100 MHz.)

4. For the same frequency, what is the nominal attenuation in dB/m?

Using a short piece of unterminated coaxial cable, connect the pulse generator to the oscilloscope vertical input via a BNC "T" connector. Figure 24-1 is a conceptual diagram of the equipment setup. Set the pulse generator to have the following output:

Amplitude   = 3.0 volts
Pulse width = 100 nsec.
Period   = 10 μsec.

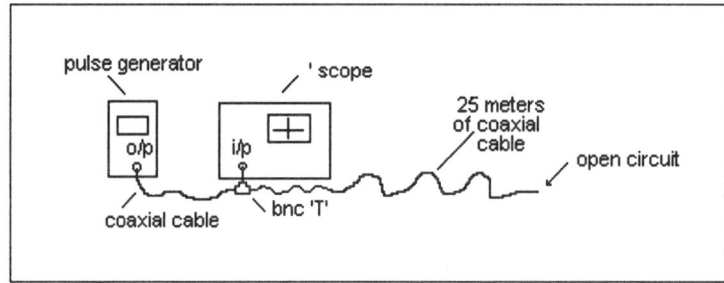

FIGURE 24-1

With the 25 meter length of open circuited coaxial cable attached to the "T" connector, adjust the time base of the oscilloscope so you can observe both the incident and reflected pulses.

5. How long does it take in nanoseconds for the pulse to travel down the line and back?

6. What is the "length" of the line in nanoseconds?

7. What is the length of the line in meters? (Assume nominal velocity of propagation found for 1.)

On the "virtual" oscilloscope screen shown below, sketch the pulse pattern *measured*. Also sketch the pulse pattern you would have *theoretically* expected. (For the theory sketch, assume no losses and use a different color.)

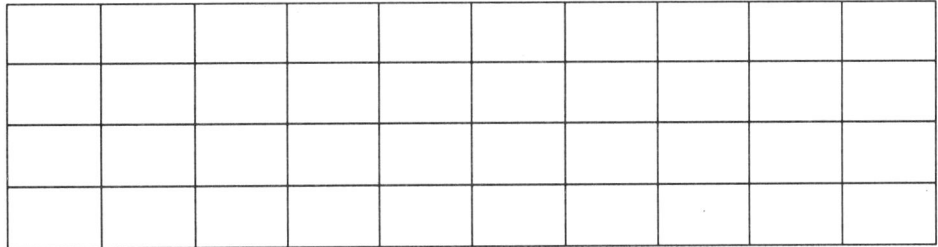

Now short out the end of the 25 meter coaxial cable and observe the result. On the oscilloscope screen below, sketch what is measured or observed. For the theoretical results, assume no losses and use a different color.

You should have noticed that the reflected pulse actually observed on the screen differs in amplitude from the incident pulse. This, of course, is due to the attenuation of the coaxial cable.

8. Using your measured results so far in the lab, calculate the attenuation in dB/m.

9. Does this calculated value compare well with the value you found in question 4? If it differs, can you explain why?

Connect various load resistors to the end of the 25 meter cable until the reflected pulse no longer appears. Measure this resistor with an ohmmeter.

10. What is the measured value of the resistor for which there is no reflected pulse?

11. What is the characteristic impedance for this cable?

Disconnect the resistor at the end of the cable and connect a 1000 pF capacitor. Make a sketch of the scope screen indicating all key voltages and times.

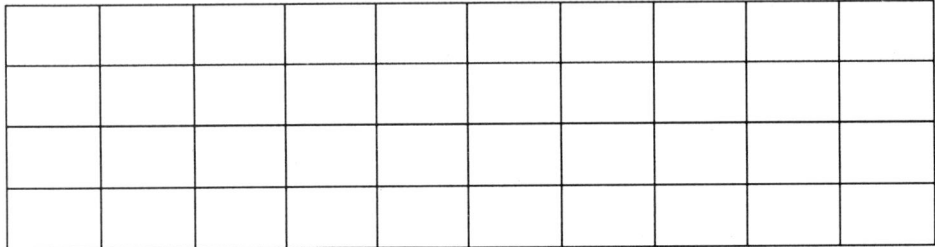

You should have observed a display that indicated something that looked like the charging of a capacitor.

Now disconnect the 25 meter section of cable and connect an open circuited section of 3 meter cable. Adjust the time base of the oscilloscope so that both the incident and reflected pulses may be observed on the screen.

12. From the measured time delay, calculate the length in meters of the cable.

On the display provided, sketch the observed pulses.

Short out the cable and observe the result. Sketch the result on the display below.

SATELLITE LINK MEASUREMENTS

LAB 25

Name: _____ Date: _____

OBJECTIVES:

Upon completion of this lab, you will be able to:

- Determine satellite link parameters
- Measure the carrier-to-noise ratio at the IF input to a satellite receiver
- Calculate and apply corrections in measuring a carrier-to-noise ratio

TEST EQUIPMENT:

- Ground level and safely accessible C-Band satellite receiving system
- Spectrum analyzer and video monitor for analyzer display to classroom
- Compass, inclinometer, measuring tape

Prerequisites

To perform this lab, the student should have received instruction on the following topics:

- Radio wave propagation
- Basic antenna concepts such as gain and beamwidth
- Fundamentals of frequency modulation
- Noise theory (noise figure, noise temperature)
- Spectral composition of a C-Band satellite downlink
- Spectrum analyzers and noise measurements

Introduction

The main objective in this lab is to analyze a satellite down link and the factors that affect the quality of the received signal from a satellite. To do this, you must have access to a commonly available C-Band television receive only setup. Since there is most likely only one receiving station setup, the lab could be performed as a group activity with instructor supervision.

When a C-Band satellite receiver demodulates the signal from a satellite transponder, the key determining factor as to picture quality on a TV set is the carrier-to-noise ratio at the satellite receiver input. The carrier-to-noise ratio in turn is set by the following set of factors or "link" parameters:

1. Effective Isotropic Radiated Power (EIRP) of the satellite transponder
2. Ratio of receiver antenna gain to receiving system noise temperature

3. Losses in the satellite signal traveling to the receiving ground station from the satellite transponder

4. Bandwidth of the transponder signal being demodulated

Some of these link parameters can be determined from simple measurements on the receiving system.

Pre-Lab

In this lab you will take a closer look at where a satellite dish has to be pointed to receive signals from a particular satellite.

In pointing the parabolic antenna of a satellite receiver at a geostationary satellite, two quantities are used in the technical description of where to point the dish.

These are the *azimuth* and the *elevation*.

1. From your text or class notes, find what is meant by the terms azimuth and elevation. Include a sketch to clarify your answer.

To locate the azimuth and elevation for a given satellite, the longitude and latitude of both the receiving ground station and satellite must be known. For example, if our ground station is located in Winnipeg and we were interested in the satellite ANIK E, then the entries in the following table would be needed.

Ground Station is at Winnipeg		Satellite is ANIK E over the equator	
LONGITUDE	LATITUDE	LONGITUDE	LATITUDE
97° 14′	49° 54′	107° 18′	0° 00′

Once we have this information, we can find out where we must point our dish in order to receive signals from the satellite by using the following formula:

azimuth = 180° + arcsin (sin ($Z-Y$) /sinθ)

where

θ = arccos (cos$X$) (cos ($Z-Y$))

$$\text{elevation} = \arctan \frac{[\cos(Z-Y)](\cos X) - 0.15126}{\sqrt{[\sin^2(Z-Y)] + [\cos^2(Z-Y)](\sin^2 X)}}$$

where

    X = Earth-site latitude (degrees)

    Y = Earth-site longitude (degrees)

    Z = Satellite longitude (degrees)

Working with these equations will give us the directions that our dish must point in order to receive signals from the satellite we have chosen.

Before you start calculating, however, you must find the longitude and latitude of *your* ground station and the longitude of the satellite *you* wish to point at. The longitude and latitude of your ground station can be found on any map of your region, and the satellite longitude should be found in your text or supplied by your instructor.

2. Calculate the azimuth and elevation for your situation. Show your work and put the finished calculations in the attached table.

Ground Station is at		Satellite is		"Look Angle" of satellite	
LONGITUDE	LATITUDE	LONGITUDE	LATITUDE	ELEVATION	AZIMUTH
			0°		

Procedure

Dish and Site Measurements

Proceed with the dish measurements (1) if the dish is ground level and safely accessible and (2) with the supervision of the instructor.

Set the controls for your satellite receiving system so that it is told to point at the satellite for which you made elevation and azimuth calculations. Adjust the channel selector until a clear unscrambled image is obtained.

For this part of the lab, we will need a compass, inclinometer, and tape measure and we will have to go outside to be at the dish itself.

Using the inclinometer, measure the elevation of the dish. Using the compass, make a rough estimate of the azimuth for the dish. The measurements with the compass will be necessarily somewhat rough. Compensate if you know the local correction for the magnetic declination.

3. What value did you measure for the elevation and how well did it compare with the calculated value?

4. What value did you measure for the azimuth and how well did it compare with the calculated value?

While you are at the dish, have the instructor point out where the polarizer is.

5. Make a sketch showing where the polarizer is. What is the purpose of the polarizer?

An important component of the satellite down link is the gain of the receiving antenna. We could make an estimate of the gain of the dish if we knew its diameter. This information will be useful to us later when we do a link analysis.

6. Using the tape measure, measure the diameter of your satellite dish.

If we have the diameter of the dish, we can calculate the dish gain for any frequency. The formula for dish gain is:

$$g = 0.55 \left(\frac{(\pi)(D)}{\lambda} \right)^2$$

$G_{dB} = 10 \log(g)$

Since the gain formula requires a wavelength, let us assume a frequency in the *center of the C band downlink* which extends from 3.7 GHz to 4.2 GHz.

7. Using the measured dish size and the assumed frequency, calculate the gain for the C band dish both in absolute terms and in dB. Show your work.

Located at the feed of the dish is a low noise amplifier (LNA). Your instructor will point it out. The specifications of this amplifier are very important in influencing the performance of the satellite receiving station. One of the more important specifications is the "temperature" of the low noise amplifier. The "temperature" rating of a satellite receiver low noise amplifier is often quoted in Kelvin degrees (K°—not °K) and has nothing to do with how hot or cold the amplifier might be but rather how much electrical noise the amplifier contributes to the receiving system. Since this amplifier is at the beginning of the receiving system and its noise output will be amplified by all following amplifiers, it is important that its noise contribution be as small as possible or that it be a "cool" amplifier—hence the term low noise amplifier or LNA. Electrical noise in TV reception shows up as "snow" on the picture or as crackle in the audio. Often the LNA temperature rating will be stamped on the amplifier or will be a stated specification with the equipment manual.

8. Examine the LNA at the feed for a temperature stamped or printed on the LNA, and if available find out the temperature rating in K°.

If you cannot find out the temperature rating by examining the feed LNA, perhaps it is available in the equipment manual. Typical LNA ratings can vary anywhere from 100 K° on down to 15 K°. (The more money you pay, the lower the rating. Also, the more recently your equipment was bought, the lower the rating, since the technology of making LNAs improves as time goes on.) If no value is available, assume 100 K° for your LNA.

Since the temperature of the low noise amplifier is related to the electrical noisiness of the amplifier, it should not be surprising that there is a relationship between noise temperature and noise figure. Given the noise figure of an amplifier, the noise temperature can be calculated and vice versa. The formula connecting the two quantities is:

Noise Figure in dB = 10 log (1 + T_{LNA}/290)

9. Calculate the noise figure for your low noise amplifier in 8.

Another parameter in determining the quality of a satellite link is the noisiness of the antenna or its "temperature." When the dish of a satellite receiving system points into the sky, some of the background radiation from the sky finds its way into the receiver. The amount of noise that does so depends on the elevation of the dish. If the dish points straight up, the "temperature" of the dish drops. If the dish is pointed at the earth, the "temperature" of the dish climbs to about 290 K°, which is due to the radiation from the earth itself. Also, the amount of electrical noise picked up varies with the size of the dish. Since the dish temperature depends on several variables, it is very site specific. Your instructor will provide you with this information.

10. What is the value of the dish temperature in K°.

Typical values for the noise temperature of a C-Band receiving antenna should normally be from 25 K° to 50 K°.

Ground Station Figure of Merit (G/T_{SYS})

As mentioned at the start, one of the factors determining the carrier-to-noise ratio at the input to the satellite receiver's demodulator is the ratio of receiving antenna gain to receiving system temperature. The system temperature in a properly designed ground station is simply the sum of the antenna temperature and low noise amplifier temperature.

$$T_{SYS} = T_{LNA} + T_{ANTENNA}$$

EXAMPLE 1: If the antenna temperature were 60 K° and the low noise amplifier temperature were 100 K°, then the system temperature would be 160 K°.

EXAMPLE 2: If the absolute antenna gain were 16000 and the system temperature were 160 K°, then the gain-to-temperature ratio for the receiving system would be 16000/160 = 100. The gain to temperature ratio for a system is almost always quoted in dB.

EXAMPLE 3: If the absolute system gain-to-temperature ratio were 100, the value in dB would be 10 log 100 = 20 dB.

---

11. Calculate or estimate the system temperature in K° for your receiving system.

12. Calculate the antenna gain to system temperature ratio in dB for your receiving system.

The reason that antenna gain to system temperature is considered a figure of merit in evaluating a satellite receiving setup is that of all the link parameters that determine the carrier-to-noise ratio, it is the only parameter that is determined by the ground station.

Free Space Loss

When the satellite signal travels from the satellite to your ground receiving station, the signal loses power depending on the signal frequency and the distance or *slant range* that it has to travel. Clearly this loss of power will affect the carrier-to-noise ratio that we can expect to have. The slant range from a particular satellite to a specific ground station can be predicted if the longitude and latitude of both the satellite and the ground receiving station are known. In that case the slant range can be calculated from the following equation.

$$S = \sqrt{(r+h)^2 + r^2 - 2(r+h)(r)(\cos X)[\cos(Z-Y)]}$$

where

S = slant range in km
X = Earth-site latitude (degrees)
Y = Earth-site longitude (degrees)
Z = Satellite longitude (degrees)
r = radius of earth = 6370 km
h = radius of geosynchronous orbit = 35,840 km

13. Using the values of longitude and latitude found previously, calculate the slant range in kM to your satellite.

Once we have the slant range or distance that the signal has to travel, we can make a reasonable estimate of the free space loss. The formula for free space loss expressed as a dB loss is:

Free Space Loss in dB = 20 log $(4\pi S/\lambda)$

where

    S is the slant range

    λ is the signal wavelength

14. Calculate the free space loss in dB for your setup. Use a frequency of 3950 MHz.

Satellite Carrier Signal Observation

Discussion of C-Band Signals

The signal frequencies present in this C-Band downlink signal range from 3700 MHz to 4200 MHz, a total of 500 MHz. The C-Band satellite has 12 transponders, each of which is allocated a bandwidth of 40 MHz. The product of 12 × 40 is 480 MHz which is how the 500 MHz is arrived at (the extra 20 MHz is used for telemetry to control the satellite). To increase the number of transponders or individual channels possible, one set of 12 transponders operates with a vertical polarization and another set of 12 transponders operates with a horizontal polarization, thus effectively doubling the channel capacity of the satellite. (In order that your receiver can discriminate between signals of vertical and horizontal polarization, there is a polarizer at the focus as you noticed earlier.)

When the signal from the satellite reaches the receiver's parabolic, dish it is brought to a focus where, as we found, there is a low noise amplifier. Next the signal is downconverted at the dish and the downconverted signal is sent down a coaxial cable to the receiver inside your lab. C-Band satellite receivers use either a single conversion technique where the signal at the feed is downconverted to 70 MHz prior to being demodulated or, as is more common, a block conversion method where the entire 500 MHz block of signals is downconverted first to an intermediate range of either 950–1450 MHz or 440–940 MHz.

The entire 500 MHz block is sent to the receiver, where there is a second down conversion to 70 MHz for each transponder signal followed by demodulation.

Observing C-Band Satellite Carrier Signals

This measurement is to be performed by the instructor as a demonstration to the class.

In this part of the lab the instructor will operate the spectrum analyzer to look at the carrier signals at the IF input to the satellite receiver electronics. The analyzer video output could be connected to the video input of a display video monitor (if one is available) for classroom demonstration and interpretation. The spectrum analyzer is connected at the down converted output and in parallel to the IF input to the satellite receiver as shown in Figure 25-1 (first check that there is no dc voltage on the signal line that could damage the spectrum analyzer before proceeding with this measurement. Also note that Figure 25-1 is a conceptual diagram only. The instructor will provide you with a diagram showing the specific details of the particular connection details for the setup in this lab). Adjust the channel selector for a clear unscrambled image on the satellite receiver television monitor.

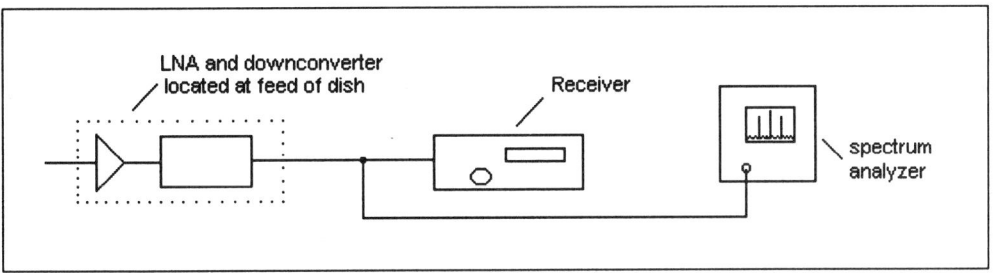

FIGURE 25-1

Remember you should see a maximum of 12 carrier signals (some transponders on the satellite may not be working, so you might see less than 12), each spaced roughly 40 MHz apart. Of course, if you change the polarizer switch setting you will see a totally different set of carriers.

Make a sketch of the analyzer display on the diagram of a spectrum analyzer screen provided below. Indicate on your sketch all key analyzer settings such as reference level value, resolution bandwidth, vertical scale factor, and center frequency. Make a rough attempt at sketching in the background thermal noise.

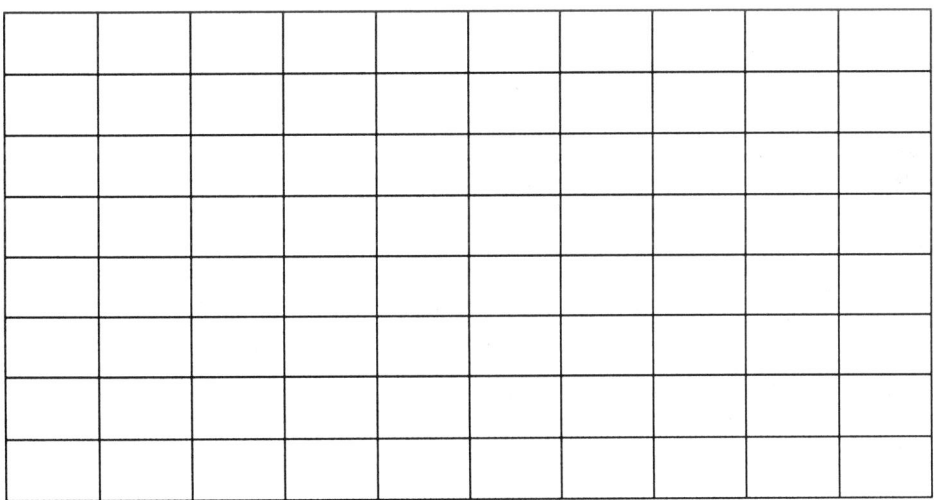

The instructor will resolve in on one of the stronger transponder signals by use of the resolution bandwidth and center frequency controls of the analyzer. You should be able to roughly estimate the carrier strength in dBm as displayed on the analyzer and make a rough guess at the average thermal noise, also in dBm, as displayed when averaged out using the video filter.

15. What is the displayed carrier strength in dBm?

16. What is the average background thermal noise in dBm? (Don't forget to add +2.5 dB to the displayed value to correct for the inaccuracies due to the analyzer's detector and log amplifier.)

17. What is the difference in dB between the displayed carrier strength (dBm) and thermal noise level (dBm)?

From these estimates it *seems* that we have made a ball park measurement of the carrier-to-noise ratio at this point in the link, but the carrier-to noise ratio we want to measure must be for the 40 MHz channel that the signal occupies. However, we are looking at the signal and its noise with an instrument—the spectrum analyzer—whose resolution bandwidth is set for a value far less than 40 MHz. As you know from noise theory, the noise power that an instrument such as an analyzer will detect goes down as the bandwidth of the instrument goes down. So what we are seeing on the analyzer screen is less thermal noise than would actually be seen. We can, however, make corrections to our measurements to compensate for this. First we have to know the analyzer bandwidth.

18. What is the resolution bandwidth setting of the analyzer?

The following example should illustrate how we can calculate a correction factor with this information.

EXAMPLE: A spectrum analyzer with a resolution bandwidth setting of 1 MHz measured the carrier-to-noise ratio to be 26 dB. The signal spectrum is spread across 40 MHz of bandwidth. What is the noise power bandwidth of the analyzer? What is the corrected value for this carrier-to-noise ratio?

Noise power bandwidth = 1.2 × 1 MHz = 1.2 MHz

Correction factor = 10 log (40 MHz/1.2 MHz) = 15.23 dB

Corrected carrier-to-noise ratio = 26 dB − 15.23 dB = 10.77 dB

---

19. Calculate your correction factor for the measurement.

20. Calculate the corrected carrier-to-noise ratio for your measurement.

Link Equation

The carrier-to-noise ratio at the receiver's demodulator input can be predicted from the following equation. (The carrier-to-noise measurement just finished was not made at the demodulator input.)

$$\frac{C}{N} = EIRP + \frac{G}{T_{SYS}} - LOSSES + 228.6 - B$$

where

EIRP is in dBW

G/T_{SYS} is in dB

LOSSES are in dB and include free space loss

B is 10 log (bandwidth)

For the above equation we know or could estimate all of the values except the EIRP. As a research project, attempt to locate a typical value for a transponder on the satellite being viewed.

21. What is a typical value or range for the EIRP of a transponder signal on the satellite being viewed?

The EIRP value that you are looking for can vary from transponder to transponder and from satellite to satellite. It is also affected by the location of your ground station. Attempt to locate a satellite footprint map for your satellite. This is essentially an EIRP contour diagram that would guide you in estimating what a typical tranponder EIRP of the satellite should be in your location.

22. If you manage to obtain an estimated EIRP, use the formula given above to calculate the carrier-to-noise expected at the receiver demodulator.

Receiver Baseband Signal

One of the output signals available from the satellite receiver is the video signal. This is roughly a 4.2 MHz video signal, as well as several other subcarriers residing between 5.8 MHz and 7.2 MHz. The word video is used because it carries the picture information. The subcarriers contain other information such as the audio information and data services information. When this baseband signal modulates the transponder carrier, it causes a maximum 10.75 MHz of deviation. By using Carson's rule (from FM theory), we can roughly estimate the required transponder bandwidth to be:

Bandwidth = 2 { $f_{MOD} + \Delta f$ } = 2 { 7.2 + 10.75 } = 35.9 MHz

The 40 MHz allocated per transponder is based on this, with 4 MHz allocated as a guardband.

The instructor will connect the baseband signal to the spectrum analyzer. Observe the video signal on the spectrum analyzer. Estimate the bandwidth.

23. What bandwidth did you observe?

SYSTEM RISE TIME AND DIGITAL TRANSMISSION

LAB 26

Name: _____   Date: _____

OBJECTIVES:

Upon completion of this lab, you will be able to:

- Measure the rise time of a pulse
- Calculate rise time of an analog circuit from measured bandwidth
- Measure and predict the effect of rise time on digital transmission

TEST EQUIPMENT:

- Function generator
- 10 Ω resistor and 0.022 µF capacitor
- Dual-trace oscilloscope

Prerequisites

To perform this lab, the student should have received instruction on the following topics:

- Pulse waveforms and pulse rise time
- The relationship between system rise time and system bandwidth
- The relationship between total system rise time and system component rise time

Procedure

Pulse Rise Time and Analog Bandwidth

An RC network will be used to model the effects of rise time on data transmission. Figure 26-1 is a conceptual diagram of the equipment and circuit setup.

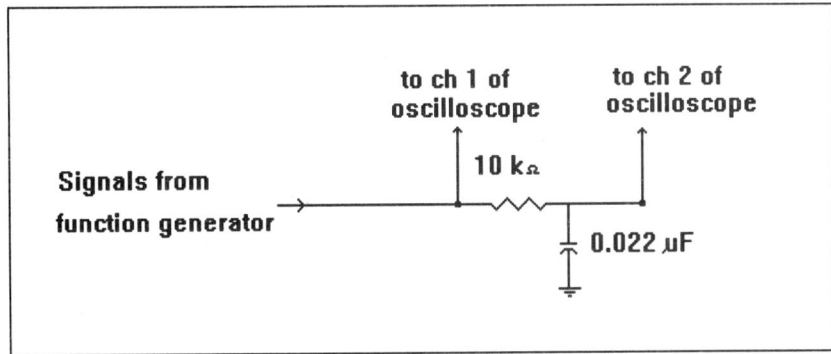

FIGURE 26-1

Measure the bandwidth of this circuit in the following manner.

With the function generator connected to the circuit input, adjust the function generator for a 100 Hz sine wave output of 10 volts peak-to-peak. Monitor and maintain this voltage level at the input, using channel 1 of the oscilloscope.

Monitor the output voltage across the capacitor with channel 2 of the oscilloscope and increase the frequency of the function generator until the output voltage falls to 7.07 volts peak-to-peak. Note this frequency, as it is the bandwidth of your circuit.

1. What is the measured bandwidth?

2. What is the theoretical bandwidth?

In calculating the theoretical bandwidth, notice that this is a low pass filter. From basic circuit theory, the bandwidth of such a circuit is given by the formula:

$$BW = \frac{1}{2\pi RC}$$

where

R = resistance in ohms

and

C = capacitance in farads

You should have calculated a value close to 723 Hz. The measured value should be close to this answer, but expect some deviation due to component tolerances.

Measure the rise time of the RC network in the following manner.

With the generator connected to the input of the circuit, adjust the function generator for a square wave output swinging from 0 volts to +10 volts at a frequency of 1 kHz. Monitor this signal on channel 1 of the oscilloscope.

Monitor the output voltage across the capacitor with channel 2 of the oscilloscope. Determine the rise time by using the following procedure.

Adjust the time base of the scope to give you the maximum expanded view of the rise time area of the pulse. Use the "var" control on the volts/div control to make the pulse fit vertically between the dashed horizontal lines marked 0% and 100% on the display. The 10% and 90% lines for the time base are marked on the display. Measure the time between the 10% and 90% lines. This is the rise time.

3. What is the measured rise time (t_r) of the network?

The measured rise time is actually the resultant rise time of the oscilloscope, the function generator, and the RC network. Since the rise times of the oscilloscope and function generator are very much smaller than that of the RC network, the effects of the scope and generator on the measured result can be ignored.

To confirm this, take a measurement of the observed rise time of the function generator connected only to the scope.

4. What is the rise time without the network?

The theoretical rise time of the RC network is related to the bandwidth of the RC circuit by the relationship:

rise time = t_r = 0.35/BW

5. Using this equation, what is the predicted rise time?

Digital Systems

In analyzing digital transmission on networks such as a fiber optic system, rise time is critical in determining the success a receiver has in detecting if a newly arrived pulse is a logic one or a logic zero.

To see why, proceed as follows using the RC network to simulate a digital link.

Adjust the function generator for a square wave output swinging from 0 volts to +10 volts at a frequency of 1 kHz with the generator connected to the input of the circuit. Monitor this signal on channel 1 of the oscilloscope.

Monitor the output voltage across the capacitor with channel 2 of the oscilloscope.

Increase the frequency of the function generator to 10 kHz and upwards and pay attention to the voltage amplitude at the output of the RC network.

6. As the frequency went up, what happened to the amplitude of the observed signal? What is the amplitude in volts at 100 kHz?

You should have observed that the amplitude of the signal became quite small. This is due to the rise time of the circuit being much larger than the width of a pulse from the generator. As a consequence the pulse never has a chance to rise to the full potential voltage of 10 volts. If the output of the RC network had to drive a circuit that needed a threshold voltage of +7 volts to detect a pulse, it is possible the voltage received would be below the needed threshold.

The result of these measurements is to bring into focus the fact that the maximum data rate on a digital network is determined by the system rise time.

Some deterioration in the rise and fall time of data is bound to take place. For this reason, a criterion has to be established to determine what is acceptable.

One possible criterion is to have the rise time of a system not be greater than 70% of the pulse width.

7. If you assumed a criterion for your RC network that the rise time could never be greater than 70% of the pulse width, what is the highest data rate that you could transmit?

8. Check your predictions with a measurement. Did the measured value agree with the theoretical value?

MANUFACTURER'S DATA SHEETS FOR LAB 10

APPENDIX A

500 Hz to 3.5 GHz

Mini-Circuits
ULTRA-REL™ MIXERS
5-YR. GUARANTEE

TAK   TSM   SAM   ROSE

	MODEL NO.	FREQUENCY MHz LO/RF f_L–f_U	FREQUENCY MHz IF	CONVERSION LOSS dB Mid-Band $\bar{x}$	CONVERSION LOSS dB Mid-Band σ	CONVERSION LOSS dB Mid-Band Max.	CONVERSION LOSS dB Total Range Max.	LO–RF ISOLATION, dB L Typ.	LO–RF ISOLATION, dB L Min.	LO–RF ISOLATION, dB M Typ.	LO–RF ISOLATION, dB M Min.	LO–RF ISOLATION, dB U Typ.	LO–RF ISOLATION, dB U Min.	LO–IF ISOLATION, dB L Typ.	LO–IF ISOLATION, dB L Min.	LO–IF ISOLATION, dB M Typ.	LO–IF ISOLATION, dB M Min.	LO–IF ISOLATION, dB U Typ.	LO–IF ISOLATION, dB U Min.	PRICE $ Qty. (1–9)	DISTRIBUTOR FACTORY	DISTRIBUTOR LOCAL
TAK case A04	TAK-5	.01–250	DC–250	4.65	.02	7.0	8.5	60	50	50	35	40	35	55	45	45	30	35	25	20.95	●	●
	TAK-5R	.05–200	DC–200	4.70	.05	6.5	8.0	55	50	45	35	45	35	50	45	40	30	40	30	20.95	●	●
	TAK-6	5–600	DC–600	5.58	.04	7.5	8.5	60	50	50	30	40	25	55	45	45	30	30	20	20.95	●	●
	TAK-6R	0.5–600	DC–600	5.40	.11	7.0	8.0	55	50	45	30	35	30	45	40	40	25	30	25	20.95	●	●
	TAK-7	2–1000	5–500	5.86	.08	7.5	8.5	45	30	35	20	30	20	45	30	35	20	30	20	20.95	●	●
TSM case A11	TSM-1	1–600	DC–600	5.71	.04	7.5	8.5	60	45	45	35	35	25	55	45	40	30	35	25	19.95	●	●
	TSM-2	1–1000	DC–1000	5.55	.08	7.5	10.0	55	45	40	20	35	18	50	40	30	25	18		20.95	●	●
	TSM-3	0.1–500	DC–500	4.75	.04	7.5	8.5	60	50	50	35	35	25	55	45	45	30	35	25	21.95	●	●
	†TSM-5	5–1500	DC–1000	6.16	.04	8.5	9.5	60	45	35	25	30	25	60	45	35	25	30	16	27.95	●	●
SAM case A03	SAM-1	1–600	DC–600	5.67	.05	7.0	8.5	55	45	45	30	35	20	50	40	40	25	30	20	16.95	●	●
	SAM-2	1–1000	DC–1000	5.68	.08	7.5	9.5	55	45	40	25	35	20	50	40	30	25	30	25	20.45	●	●
	SAM-3	0.1–500	DC–500	5.04	.07	7.0	8.5	60	50	50	35	35	30	50	40	45	30	30	20	19.95	●	●
	SAM-4	5–1250	0.5–600	5.98	.18	8.5	8.5	55	40	35	25	30	20	45	35	25	30	20		23.95	●	●
	SAM-5	5–1500	0.5–1000	5.81	.08	7.5	8.5	55	40	35	25	30	20	50	40	35	25	30	20	29.95	●	●
ROSE case PP94	ROSE-1	1–600	DC–600	5.08	.03	6.5	7.5	40	30	35	25	30	20	55	40	40	20	25	18	11.95	●	●
	ROSE-2	1–1000	DC–1000	5.60	.23	7.0	8.0	61	45	37	22	25	18	55	40	26	17	16	12	16.95	●	●

L = low range (f_L to 10 f_L)   M = mid range (10 f_L to $f_U/2$)   U = upper range ($f_U/2$ to f_U)
m = mid band (2 f_L to $f_U/2$)

schematic

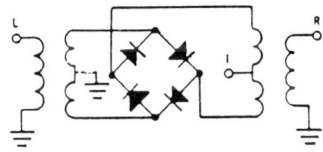

MIL-M-28837/1, NSN GUIDE

MCL NO.	NSN	MIL-M-28837/1*
GRA-1	5895-00-480-2849	
GRA-3	5895-01-169-1815	
GRA-8	5895-01-217-5627	
SAM-1	5895-01-117-2926	
SAM-2	5985-01-165-6621	
SAM-3	5895-01-062-9973	
SAM-5	5895-01-036-9507	
SBL-1	5895-01-126-4913	
SBL-1X	5895-01-179-8084	
SRA-1	5895-00-008-8272	03
SRA-1-1	5962-01-113-5431	
SRA-1-TX	5895-01-163-9247	
SRA-1-1-TX	5895-01-151-6753	
SRA-11-TX	5895-01-163-9248	
SRA-1W	5895-01-163-0433	09
SRA-3	5895-01-021-5914	
SRA-4	5826-01-155-6545	
SRA-6	5895-01-124-0117	
SRA-8	5985-01-081-0977	
SRA-11	5895-01-273-0883	
SBL-3	5895-01-326-6030	
TAK-5	5895-01-271-0842	
TAK-6	5895-01-231-2372	
TSM-1	5895-01-121-7958	

*units are not QPL-listed

pin connections see case style outline drawing

TAK				TSM	SAM			ROSE
–5 –6	–7	–5R	–6R	all models (△)	–1 –3	–4 –5	–2	–1
8	8	8	1	8	8	8	8	1
1	3,4*	1	8	1	1	3,4*	1	3
3,4*	1	3,4*	5,6*	3,4*	3,4*	1	3,4*	2
2,5,6,7	2,5,6,7	2,5,6,7	2,3,4,7	2,5,6,7	2,5,6,7	2,5,6,7	—	
2	2,5,6,7	—	3,4,7	2	2		2,5,6,7	4

△ TSM-5, CASE GROUND 2,7; TSM-2 CASE GROUND 2, 5, 6, 7.
* Pins must be connected together externally.

In Stock...Immediate Delivery

Reprinted with permission of Mini-Circuits.

most economical
Frequency Mixers Models
SBL-1
LEVEL 7 (+7dBm LO, up to +1dBm RF)

computer-automated performance data
typical production unit / for data of other models consult factory

mixer conversion loss and isolation

RF MHz	LO MHz	Conversion Loss (dB)			Isolation L-R (dB)			Isolation L-I (dB)		
		LO +4dBm	LO +7dBm	LO +10dBm	LO +4dBm	LO +7dBm	LO +10dBm	LO +4dBm	LO +7dBm	LO +10dBm
1.000	31.00	6.99	6.67	6.45	62.73	>67.00	68.01	61.88	65.61	>70.00
2.000	32.00	6.50	6.24	6.05	61.86	>67.00	67.46	61.93	>67.00	>70.00
5.000	35.00	5.96	5.74	5.60	61.73	64.84	66.12	62.43	>67.00	>70.00
10.000	40.00	5.85	5.58	5.46	61.35	64.19	65.15	61.72	64.81	66.04
20.000	50.00	5.96	5.67	5.51	60.14	62.22	62.01	60.33	61.69	61.30
32.188	62.19	5.86	5.60	5.51	58.85	59.04	58.99	58.27	57.74	57.09
50.000	80.00	5.84	5.60	5.49	56.98	56.71	55.75	54.84	54.39	53.61
78.970	48.97	5.83	5.56	5.44	54.05	52.21	50.75	51.16	49.67	48.53
100.000	70.00	5.74	5.52	5.43	51.39	49.41	48.33	48.54	46.90	46.10
156.940	126.94	5.69	5.53	5.47	45.18	44.20	43.99	42.22	41.66	41.56
200.000	170.00	5.85	5.68	5.60	41.88	41.56	41.75	39.00	38.98	39.65
203.720	173.72	5.83	5.67	5.56	41.46	41.05	41.34	38.60	38.52	39.23
250.500	220.50	5.80	5.63	5.47	40.14	40.04	40.21	37.21	37.29	37.77
297.290	267.29	5.77	5.61	5.48	36.89	36.90	37.90	33.86	33.97	34.98
344.070	314.07	5.93	5.81	5.72	36.80	37.11	37.44	33.43	33.88	34.36
375.260	345.26	6.28	6.13	5.97	34.83	36.02	37.25	31.69	32.79	34.20
406.440	376.44	6.27	6.15	6.07	33.09	34.80	36.66	29.68	31.05	32.79
437.630	407.63	6.38	6.08	5.88	32.48	33.79	35.38	29.38	30.39	31.81
468.820	438.82	6.48	6.00	5.73	32.24	32.80	34.05	29.76	30.12	30.96
500.000	470.00	6.90	6.31	5.99	32.48	32.36	33.04	30.19	30.37	30.65

mixer VSWR   φ detection

freq. MHz	VSWR, RF port			VSWR, LO port			VSWR, IF port			Freq. (MHz)	max. DC output mV	L-C offset mV
	LO +4dBm	LO +7dBm	LO +10dBm	LO +4dBm	LO +7dBm	LO +10dBm	LO +4dBm	LO +7dBm	LO +10dBm			
1.001	1.83	1.80	1.86	2.45	3.75	5.47	1.67	1.44	1.30	10.000	-233.19	+0.00
16.121	1.07	1.14	1.20	1.92	2.81	3.95	1.68	1.45	1.31	20.000	-230.80	-.01
20.000	1.06	1.14	1.20	1.94	2.89	3.95	1.68	1.44	1.31	38.000	-225.55	+0.00
31.242	1.06	1.14	1.20	1.93	2.83	3.88	1.68	1.44	1.32	50.000	-225.92	+0.00
50.000	1.07	1.15	1.20	1.83	2.69	3.54	1.69	1.46	1.33	66.000	-226.43	+0.00
76.605	1.08	1.16	1.21	1.84	2.72	3.70	1.71	1.48	1.36	80.000	-230.22	+.01
100.000	1.09	1.17	1.22	1.82	2.62	3.38	1.74	1.52	1.39	100.000	-231.28	+.03
121.970	1.11	1.12	1.24	1.82	2.62	3.53	1.79	1.55	1.42	136.000	-229.89	+.08
152.210	1.13	1.20	1.25	1.88	2.66	3.42	1.85	1.62	1.50	178.000	-223.93	+.17
182.450	1.15	1.22	1.27	1.83	2.62	3.21	1.93	1.69	1.57	200.000	-223.43	+.15
200.000	1.12	1.26	1.27	1.87	2.58	3.22	1.98	1.77	1.62	234.000	-223.86	+.23
242.940	1.19	1.26	1.32	1.89	2.56	3.19	2.11	1.88	1.74	262.000	-229.61	+.25
273.180	1.21	1.28	1.34	1.90	2.59	3.17	2.21	1.99	1.86	304.000	-218.66	+.32
318.540	1.23	1.30	1.35	1.99	2.64	3.25	2.35	2.11	1.95	360.000	-194.91	+.72
348.780	1.24	1.31	1.36	2.01	2.59	3.18	2.38	2.15	2.03	402.000	-179.58	+.97
379.030	1.24	1.30	1.34	2.03	2.60	3.15	2.45	2.24	2.09	444.000	-163.75	+.88
409.270	1.23	1.28	1.32	2.12	2.72	3.19	2.46	2.24	2.11	458.000	-160.12	+.78
424.390	1.21	1.27	1.31	2.20	2.77	3.27	2.50	2.30	2.16	472.000	-155.46	+.60
454.630	1.19	1.24	1.28	2.30	2.82	3.34	2.46	2.32	2.14	486.000	-150.45	+.39
500.000	1.16	1.21	1.25	2.40	3.01	3.54	2.43	2.23	2.12	500.000	-142.53	+.23

Measurements at RF & LO Power +7 dBm

Mini-Circuits P.O. Box 350166, Brooklyn, New York 11235-0003 (718) 934-4500 Fax (718) 332-4661
Distribution Centers NORTH AMERICA 800-654-7949 417-335-5935 Fax 417-335-5945 EUROPE 44-252-835094 Fax 44-252-837010

most widely-used
Frequency Mixers

LEVEL 7 (+7 dBm LO, up to +1 dBm RF)

case style selection
outline drawings Table of Contents

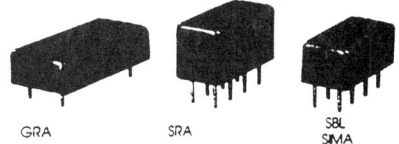

GRA   SRA   SBL SIMA

	MODEL NO.	FREQUENCY MHz LO/RF f_L–f_U	FREQUENCY MHz IF	CONVERSION LOSS dB Mid-Band m $\bar{X}$	σ	Max.	Total Range Max.	LO-RF ISOLATION, dB L Typ.	Min.	M Typ.	Min.	U Typ.	Min.	LO-IF ISOLATION, dB L Typ.	Min.	M Typ.	Min.	U Typ.	Min.	PRICE $ Qty. (1-9)	DISTRIBUTOR FACTORY	LOCAL
GRA case D09	GRA-1	5-500	DC-500	5.96	.09	7.0	8.5	50	45	45	30	35	25	45	35	40	25	30	20	14.95	●	●
	GRA-3	025-200	DC-200	4.78	.07	7.5	8.5	60	50	45	35	35	25	45	30	40	30	30	20	18.45	●	●
	GRA-6	003-100	DC-100	4.70	.14	7.5	8.5	60	50	45	30	35	25	60	45	40	25	30	20	28.95	●	●
	GRA-8	0005-10	DC-10	6.03	.08	7.5	8.5	60	50	50	40	45	35	60	50	50	40	45	35	31.95	●	●
SRA case A01	●SRA-1	.5-500	DC-500	5.11	.09	7.0	8.5	50	45	45	30	35	25	45	35	40	25	30	20	13.45	●	●
	SRA-1TX	5-500	DC-500	5.68	.04	7.0	8.5	50	45	45	30	35	25	45	35	40	25	30	20	59.95	●	●
	SRA-1W	1-750	DC-750	5.80	.04	7.5	8.5	50	45	45	30	35	25	45	30	40	25	30	20	15.95	●	●
	SRA-1-1	1-500	DC-500	4.81	.11	7.5	8.5	50	45	45	30	35	25	45	30	40	25	30	20	14.95	●	●
	SRA-2	1-1000	5-500	5.66	.07	7.5	8.5	45	30	35	20	30	15	45	30	30	20	30	20	15.95	●	●
	SRA-2CM	5-1000	DC-1000	5.27	.04	7.0	8.5	60	50	35	30	30	25	50	45	30	25	25	20	14.95	●	●
	SRA-3	025-200	DC-200	4.61	.06	7.5	8.5	60	50	45	35	35	25	45	35	40	30	30	20	15.95	●	●
	SRA-4	5-1250	5-500	5.71	.08	7.5	8.5	50	40	40	20	30	20	50	40	40	20	30	20	17.95	●	●
	*SRA-5	5-1500	10-600	6.69	.07	8.0	8.5	50	45	35	30	30	20	45	40	30	25	25	15	24.95	●	●
	SRA-6	003-100	DC-100	4.58	.05	7.5	8.5	60	50	45	30	35	25	60	45	40	25	30	20	24.95	●	●
	SRA-8	0005-10	DC-10	5.69	.11	7.5	8.5	60	50	50	40	45	35	60	50	50	40	45	35	29.95	●	●
case A06	SRA-11	5-2000	10-600	6.72	.07	8.5	9.0	50	45	35	25	30	20	45	40	30	20	25	15	20.95	●	●
	SRA-12	800-1250	50-90	6.21	.13	7.5	7.5	32	25	35	25	35	25	30	20	30	20	30	20	29.95	●	●
	SRA-149	5-500	DC-500	5.61	.07	6.5	8.0	60	50	55	45	53	40	50	40	35	25	30	24	9.95	●	●
	SRA-2000	100-2000	DC-600	8.60	.15	9.5	9.5	37 (typ.)		20 (min.)				30 (typ.)		20 (min.)				21.95	●	●
	†SRA-2400	750-2400	DC-400	5.95	.26	9.0	9.0	30	20	30	20	30	20	30	8	30	8	30	8	22.95	●	●
	**†SRA-3500	500-3500	DC-1000	7.28	.31	9.5	9.5	30	17	30	17	30	17	20	8	20	8	20	8	28.95	●	●
□SBL case A06	SBL-1	1-500	DC-500	5.60	.09	7.0	8.0	60	45	45	35	40	25	45	35	40	25	30	20	4.75	●	●
	SBL-1X	10-1000	5-500	5.88	.10	7.5	8.0	50	40	40	30	30	20	50	45	40	35	35	25	6.45	●	●
	SBL-1Z	10-1000	DC-500	6.27	.09	7.5	9.0	50	40	35	25	25	20	40	25	25	18	19	15	7.45	●	●
	SBL-1-1	0.1-400	DC-400	4.84	.04	7.0	8.0	50	45	45	30	35	25	45	30	40	25	30	20	7.45	●	●
	SBL-3	025-200	DC-200	4.81	.05	7.5	8.5	55	50	45	30	35	25	45	35	40	30	30	20	7.45	●	●
	○SBL-11	5-2000	10-600	7.08	.11	8.5	9.0	50	45	35	25	30	20	45	40	30	20	25	15	19.95	●	●
SIMA case A06	SIMA-5	2-1500	DC-1000	7.01	.08	8.0	9.0	65	44	44	23	31	22	54	38	30	18	25	11	24.95	●	●

L = low range (f_L to 10 f_L)   M = mid range (10 f_L to f_U/2)   U = upper range (f_U/2 to f_U)
m = mid band (2 f_L to f_U/2)

NOTES:
† Phase Detection, Polarity Positive
* SRA-5 case style is A06.
● HTRB tested, 5-year guarantee
** Conversion loss 10 dB max. at IF = 1000 MHz.
○ Pin connections same as the SRA-11
□ NON-HERMETIC
$\bar{X}$ = Average of conversion loss at center of mid-band frequency ($f_L + f_U/4$).

σ = Standard deviation.
1. For quality control procedures, environmental specifications, and Hi-Rel, MIL and TX description see Table of Contents.
2. Absolute Maximum Ratings: RF power 50 mW, peak IF current 40 mA, see Table of Contents.
3. Prices and specifications subject to change without notice.

pin connections see case style outline drawing

Series	GRA	SRA						SBL			SIMA
models	all models	-1 -1TX -1-1 -3	-2 -4	-6 -8	-1W -2CM	-5 -11 -12 -2000 -2400 -3500	-149***	-1 -1-1 -3	-1X	-1Z	-5
LO	1	8	8	8	8	8	2	8	8	1	8
RF	6	1	3,4*	1	1	1	2	1	3,4*	8	1
IF	4	3,**	1	3,4*	3,4*	3	5,6*	3,4*	1	3	3
GND ♦	2,3,5	2,5,6,7	2,5,6,7	2,5,6,7	2,5,6,7	1		2,5,6,7	2,5,6,7	2,5,6,7	4
CASE GND	—	2	2,5,6,7	—	2,5,6,7	2,5,6,7	3,4,7	—	2,5,6,7	2,5,6,7	2,5,6,7

* Pins must be connected together externally   : LO = 1; RF = 8; IF 3
*** Blue bead is pin #4
♦ Ground externally. All measurements made with GND pin(s) grounded externally.

Mini-Circuits™
P.O. Box 350166 Brooklyn, New York 11235-0003 (718) 934-4500 Fax (718) 332-4661
Distribution Centers NORTH AMERICA 800-654-7949 417-335-5935 Fax 417-335-5945   EUROPE 44-252-835094 Fax 44-252-837010

Mini-Circuits ULTRA-REL™ MIXERS
5-YR. GUARANTEE

1 to 500 MHz

00-08

mixer harmonic intermodulation
(relative to desired IF output)

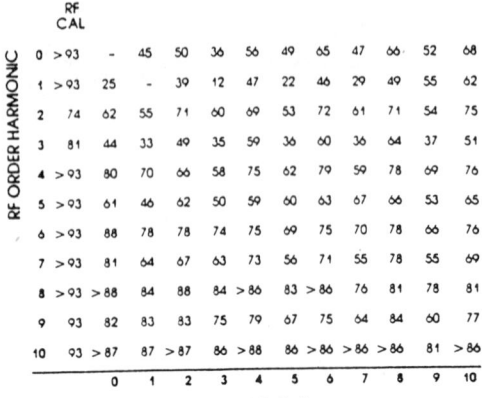

RF ORDER HARMONIC	RF CAL	0	1	2	3	4	5	6	7	8	9	10
0	>93	–	45	50	36	56	49	65	47	66	52	68
1	>93	25	–	39	12	47	22	46	29	49	55	62
2	74	62	55	71	60	69	53	72	61	71	54	75
3	81	44	33	49	35	59	36	60	36	64	37	51
4	>93	80	70	66	58	75	62	79	59	78	69	76
5	>93	61	46	62	50	59	60	63	67	66	53	65
6	>93	88	78	78	74	75	69	75	70	78	66	76
7	>93	81	64	67	63	73	56	71	55	78	55	69
8	>93	>88	84	88	84	>86	83	>86	76	81	78	81
9	93	82	83	83	75	79	67	75	64	84	60	77
10	93	>87	87	>87	86	>88	86	>86	>86	>86	81	>86

Harmonic LO Order

test conditions: RF = 185.100 MHZ INPUT P. = +0.00DBM
LO = 155.010 MHZ INPUT P. = 6.92DBM
IF = 30.090 MHZ IF AMPLIT. = -6.11 DBM

RF ORDER HARMONIC	RF CAL	0	1	2	3	4	5	6	7	8	9	10
0	>84	–	36	38	24	41	35	45	30	52	31	52
1	>84	25	–	88	12	50	19	41	21	37	40	47
2	82	63	55	68	54	66	52	64	52	66	54	74
3	>84	66	53	66	58	71	55	75	52	65	54	73
4	>84	>76	75	>76	75	>77	76	>77	75	>76	74	>76
5	>84	>76	>76	>76	>76	>76	>77	>77	73	>76	71	>76
6	>84	>78	>76	>76	>76	>76	>76	>77	>77	>77	>76	>76
7	>84	>78	>78	>76	>76	>76	>76	>76	>77	>77	>77	>76
8	>84	>78	>76	>78	>78	>76	>76	>76	>76	>77	>77	>77
9	>84	>78	>78	>78	>78	>78	>76	>76	>76	>76	>76	>77
10	>83	>77	78	>78	>78	>78	>78	>76	>76	>76	>76	>76

Harmonic LO Order

test conditions: RF = 185.100 MHZ INPUT P. = -10.03DBM
LO = 155.010 MHZ INPUT P. = +6.92DBM
IF = 30.090 MHZ IF AMPLIT. = -15.84 DBM

typical performance curves
(production unit)

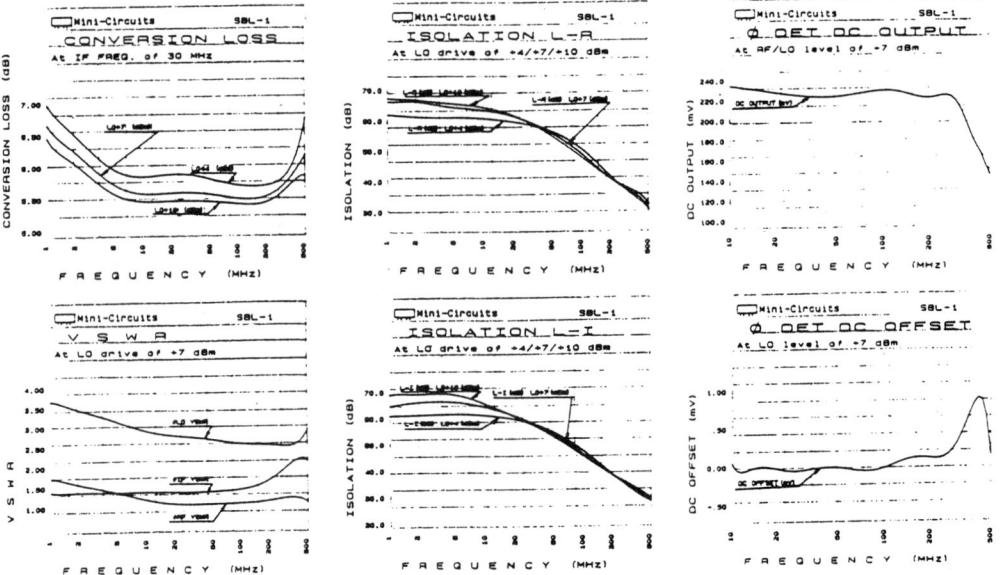

In Stock…immediate Delivery

1-63

MANUFACTURER'S DATA SHEETS FOR LAB 20

APPENDIX B

INS8250, INS8250-B Universal Asynchronous Receiver/Transmitter

General Description

Each of these parts function as a serial data input/output interface in a microcomputer system. The system software determines the functional configuration of the UART via a TRI-STATE® 8-bit bidirectional data bus.

The UART performs serial-to-parallel conversion on data characters received from a peripheral device or a MODEM, and parallel-to-serial conversion on data characters received from the CPU. The CPU can read the complete status of the UART. Status information reported includes the type and condition of the transfer operations being performed by the UART, as well as any error conditions (parity, overrun, framing, or break interrupt).

The UART includes a programmable baud rate generator that is capable of dividing the timing reference clock input by divisors of 1 to ($2^{16}-1$), and producing a 16 × clock for driving the internal transmitter logic. Provisions are also included to use this 16 × clock to drive the receiver logic. The UART includes a complete MODEM-control capability and a processor-interrupt system. Interrupts can be programmed to the user's requirements minimizing the computing required to handle the communications link.

National's INS8250 universal asynchronous receiver transmitter (UART) is the unanimous choice of almost every PC and add-on manufacturer in the world. The INS8250 is a programmable communications chip available in a standard 40-pin dual-in-line and a 44-pin PCC package. The chip is fabricated using N-channel silicon gate technology.

Features

- Easily interfaces to most popular microprocessors.
- Adds or deletes standard asynchronous communication bits (start, stop, and parity) to or from serial data stream.
- Holding and shift registers eliminate the need for precise synchronization between the CPU and the serial data.
- Independently controlled transmit, receive, line status, and data set interrupts.
- Programmable baud generator allows division of any input clock by 1 to ($2^{16}-1$) and generates the internal 16 × clock.
- Independent receiver clock input.
- MODEM control functions (CTS, RTS, DSR, DTR, RI, and DCD).
- Fully programmable serial-interface characteristics:
  — 5-, 6-, 7-, or 8-bit characters
  — Even, odd, or no-parity bit generation and detection
  — 1-, 1½-, or 2-stop bit generation
  — Baud generation (DC to 56k baud).
- False start bit detection.
- Complete status reporting capabilities.
- TRI-STATE TTL drive capabilities for bidirectional data bus and control bus.
- Line break generation and detection.
- Internal diagnostic capabilities:
  — Loopback controls for communications link fault isolation
  — Break, parity, overrun, framing error simulation.
- Fully prioritized interrupt system controls.

Connection Diagram

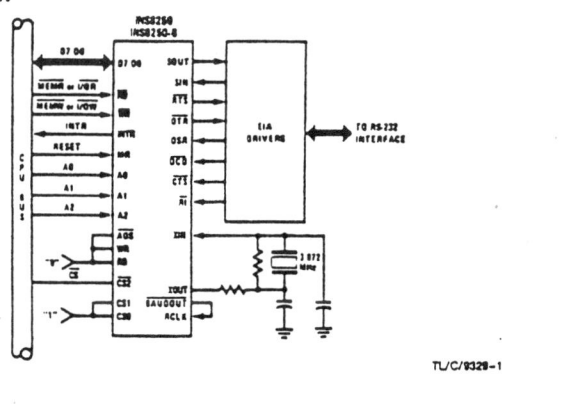

Reprinted with permission of National Semiconductor Corporation.

5.0 Block Diagram

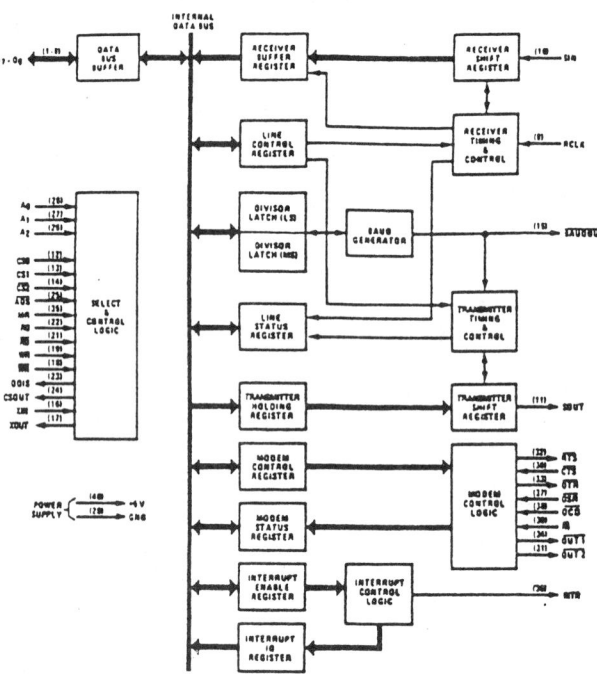

Note: Applicable pinout numbers are included within parenthesis.

6.0 Pin Descriptions

The following describes the function of all UART, pins. Some of these descriptions reference internal circuits.

In the following descriptions, a low represents a logic 0 (0V nominal) and a high represents a logic 1 (+2.4V nominal).

6.1 INPUT SIGNALS

Chip Select (CS0, CS1, $\overline{CS2}$), Pins 12–14: When CS0 and CS1 are high and $\overline{CS2}$ is low, the chip is selected. This enables communication between the UART and the CPU. The positive edge of an active Address Strobe signal latches the decoded chip select signals, completing chip selection. If $\overline{ADS}$ is always low valid chip selects should stabilize according to the t_{CSW} parameter.

Read (RD, $\overline{RD}$), Pins 22 and 21: When RD is high or $\overline{RD}$ is low while the chip is selected, the CPU can read status information or data from the selected UART register.

Note: Only an active RD or $\overline{RD}$ input is required to transfer data from the UART during a read operation. Therefore, tie either the RD input permanently low or the $\overline{RD}$ input permanently high, when it is not used.

Write (WR, $\overline{WR}$), Pins 19 and 18: When WR is high or $\overline{WR}$ is low while the chip is selected, the CPU can write control words or data into the selected UART register.

Note: Only an active WR or $\overline{WR}$ input is required to transfer data to the UART during a write operation. Therefore, tie either the WR input permanently low or the $\overline{WR}$ input permanently high, when it is not used.

Address Strobe ($\overline{ADS}$), Pin 25: The positive edge of an active Address Strobe ($\overline{ADS}$) signal latches the Register Select (A0, A1, A2) and Chip Select (CS0, CS1, CS2) signals.

Note: An active $\overline{ADS}$ input is required when the Register Select (A0, A1, A2) signals are not stable for the duration of a read or write operation. If not required, tie the $\overline{ADS}$ input permanently low.

Register Select (A0, A1, A2), Pins 26–28: Address signals connected to these 3 inputs select a UART register for the CPU to read from or write to during data transfer. A table of registers and their addresses is shown below. Note that the state of the Divisor Latch Access Bit (DLAB), which is the most significant bit of the Line Control Register, affects the selection of certain UART registers. The DLAB must be set high by the system software to access the Baud Generator Divisor Latches.

6.0 Pin Descriptions (Continued)

DLAB	A₂	A₁	A₀	Register
0	0	0	0	Receiver Buffer (read), Transmitter Holding Register (write)
0	0	0	1	Interrupt Enable
X	0	1	0	Interrupt Identification (read only)
X	0	1	1	Line Control
X	1	0	0	MODEM Control
X	1	0	1	Line Status
X	1	1	0	MODEM Status
1	0	0	0	Divisor Latch (least significant byte)
1	0	0	1	Divisor Latch (most significant byte)

Register Addresses

Master Reset (MR), Pin 35: When this input is high, it clears all the registers (except the Receiver Buffer, Transmitter Holding, and Divisor Latches), and the control logic of the UART. The states of various output signals (SOUT, INTR, $\overline{OUT\ 1}$, $\overline{OUT\ 2}$, $\overline{RTS}$, $\overline{DTR}$) are affected by an active MR input. (Refer to Table I.).

Receiver Clock (RCLK), Pin 9: This input is the 16 × baud rate clock for the receiver section of the chip.

Serial Input (SIN), Pin 10: Serial data input from the communications link (peripheral device, MODEM, or data set).

Clear to Send ($\overline{CTS}$), Pin 36: When low, this indicates that the MODEM or data set is ready to exchange data. The $\overline{CTS}$ signal is a MODEM status input whose conditions can be tested by the CPU reading bit 4 (CTS) of the MODEM Status Register. Bit 4 is the complement of the $\overline{CTS}$ signal. Bit 0 (DCTS) of the MODEM Status Register indicates whether the $\overline{CTS}$ input has changed state since the previous reading of the MODEM Status Register. $\overline{CTS}$ has no effect on the Transmitter.

Note: Whenever the CTS bit of the MODEM Status Register changes state, an interrupt is generated if the MODEM Status Interrupt is enabled.

Data Set Ready ($\overline{DSR}$), Pin 37: When low, this indicates that the MODEM or data set is ready to establish the communications link with the UART. The $\overline{DSR}$ signal is a MODEM status input whose condition can be tested by the CPU reading bit 5 (DSR) of the MODEM Status Register. Bit 5 is the complement of the $\overline{DSR}$ signal. Bit 1 (DDSR) of the MODEM Status Register indicates whether the $\overline{DSR}$ input has changed state since the previous reading of the MODEM Status Register.

Note: Whenever the DSR bit of the MODEM Status Register changes state, an interrupt is generated if the MODEM Status Interrupt is enabled.

Data Carrier Detect ($\overline{DCD}$), Pin 38: When low, indicates that the data carrier has been detected by the MODEM or data set. The $\overline{DCD}$ signal is a MODEM status input whose condition can be tested by the CPU reading bit 7 (DCD) of the MODEM Status Register. Bit 7 is the complement of the $\overline{DCD}$ signal. Bit 3 (DDCD) of the MODEM Status Register indicates whether the $\overline{DCD}$ input has changed state since the previous reading of the MODEM Status Register. $\overline{DCD}$ has no effect on the receiver.

Note: Whenever the DCD bit of the MODEM Status Register changes state, an interrupt is generated if the MODEM Status Interrupt is enabled.

Ring Indicator ($\overline{RI}$), Pin 39: When low, this indicates that a telephone ringing signal has been received by the MODEM or data set. The $\overline{RI}$ signal is a MODEM status input whose condition can be tested by the CPU reading bit 6 (RI) of the MODEM Status Register. Bit 6 is the complement of the $\overline{RI}$ signal. Bit 2 (TERI) of the MODEM Status Register indicates whether the $\overline{RI}$ input signal has changed from a low to a high state since the previous reading of the MODEM Status Register.

Note: Whenever the RI bit of the MODEM Status Register changes from a high to a low state, an interrupt is generated if the MODEM Status Interrupt is enabled.

V$_{CC}$, Pin 40: +5V supply.

V$_{SS}$, Pin 20: Ground (0V) reference.

6.2 OUTPUT SIGNALS

Data Terminal Ready ($\overline{DTR}$), Pin 33: When low, this informs the MODEM or data set that the UART is ready to establish a communications link. The $\overline{DTR}$ output signal can be set to an active low by programming bit 0 (DTR) of the MODEM Control Register to a high level. A Master Reset operation sets this signal to its inactive (high) state.

Request to Send ($\overline{RTS}$), Pin 32: When low, this informs the MODEM or data set that the UART is ready to exchange data. The $\overline{RTS}$ output signal can be set to an active low by programming bit 1 (RTS) of the MODEM Control Register. A Master Reset operation sets this signal to its inactive (high) state.

Output 1 ($\overline{OUT\ 1}$), Pin 34: This user-designated output can be set to an active low by programming bit 2 (OUT 1) of the MODEM Control Register to a high level. A Master Reset operation sets this signal to its inactive (high) state. In the XMOS parts this will achieve TTL levels.

Output 2 ($\overline{OUT\ 2}$), Pin 31: This user-designated output can be set to an active low by programming bit 3 (OUT 2) of the MODEM Control Register to a high level. A Master Reset operation sets this signal to its inactive (high) state. In the XMOS parts this will achieve TTL levels.

Chip Select Out (CSOUT), Pin 24: When high, it indicates that the chip has been selected by active, CS0, CS1, and $\overline{CS2}$ inputs. No data transfer can be initiated until the CSOUT signal is a logic 1. CSOUT goes low when the UART is deselected.

Driver Disable (DDIS), Pin 23: This goes low whenever the CPU is reading data from the UART. It can disable or control the direction of a data bus transceiver between the CPU and the UART (see Typical Interface for a High Capacity Data Bus).

Baud Out ($\overline{BAUDOUT}$), Pin 15: This is the 16 × clock signal from the transmitter section of the UART. The clock rate is equal to the main reference oscillator frequency divided by the specified divisor in the Baud Generator Divisor Latches. The $\overline{BAUDOUT}$ may also be used for the receiver section by tying this output to the RCLK input of the chip.

6.0 Pin Descriptions (Continued)

Interrupt (INTR), Pin 30: This goes high whenever any one of the following interrupt types has an active high condition and is enabled via the IER: Receiver Line Status; Received Data Available; Transmitter Holding Register Empty; and MODEM Status. The INTR signal is reset low upon the appropriate interrupt service or a Master Reset operation.

Serial Output (SOUT), Pin 11: This is the composite serial data output to the communications link (peripheral, MODEM or data set). The SOUT signal is set to the Marking (logic 1) state upon a Master Reset operation or when the transmitter is idle.

6.3 INPUT/OUTPUT SIGNALS

Data (D_7–D_0) Bus, Pins 1–8: This bus is comprised of eight TRI-STATE input/output lines. The bus provides bidirectional communications between the UART and the CPU. Data, control words, and status information are transferred via the D_7–D_0 Data Bus.

External Clock Input/Output (X_{IN}, X_{OUT}) Pins 16 and 17: These two pins connect the main timing reference (crystal or signal clock) to the UART. When a crystal oscillator or a clock signal is provided, it drives the UART via XIN (see typical oscillator network illustration).

7.0 Connection Diagrams

Dual-In-Line Package

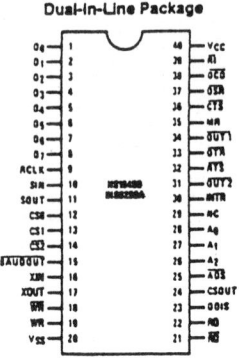

Top View

Order Number INS8250N, INS8250N-B or INS8250N/A+
See NS Package Number N40A

PCC Package

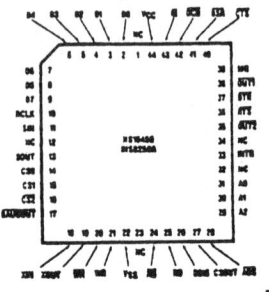

Top View

Order Number INS8250V-B
See NS Package Number V44A

TABLE I. UART Reset Functions

Register/Signal	Reset Control	Reset State
Interrupt Enable Register	Master Reset	0000 0000 (Note 1)
Interrupt Identification Register	Master Reset	0000 0001
Line Control Register	Master Reset	0000 0000
MODEM Control Register	Master Reset	0000 0000
Line Status Register	Master Reset	0110 0000
MODEM Status Register	Master Reset	XXXX 0000 (Note 2)
SOUT	Master Reset	High
INTR (RCVR Errs)	Read LSR/MR	Low
INTR (RCVR Data Ready)	Read RBR/MR	Low
INTR (THRE)	Read IIR/Write THR/MR	Low
INTR (Modem Status Changes)	Read MSR/MR	Low
OUT 2	Master Reset	High
RTS	Master Reset	High
DTR	Master Reset	High
OUT 1	Master Reset	High

Note 1: Underlined bits are permanently low.
Note 2: Bits 7–4 are driven by the input signals.

8.0 Registers

The system programmer may access any of the UART registers summarized in Table II via the CPU. These registers control UART operations including transmission and reception of data. Each register bit in Table II has its name and reset state shown.

8.1 LINE CONTROL REGISTER

The system programmer specifies the format of the asynchronous data communications exchange and sets the Divisor Latch Access bit via the Line Control Register (LCR). The programmer can also read the contents of the Line Control Register. The read capability simplifies system programming and eliminates the need for separate storage in system memory of the line characteristics. Table II shows the contents of the LCR. Details on each bit follow:

Bits 0 and 1: These two bits specify the number of bits in each transmitted or received serial character. The encoding of bits 0 and 1 is as follows:

Bit 1	Bit 0	Character Length
0	0	5 Bits
0	1	6 Bits
1	0	7 Bits
1	1	8 Bits

Bit 2: This bit specifies the number of Stop bits transmitted and received in each serial character. If bit 2 is a logic 0, one Stop bit is generated or checked in the serial data. If bit 2 is a logic 1 when a 5-bit word length is selected via bits 0

TABLE II. Summary of Registers

Bit No.	0 DLAB=0 Receiver Buffer Register (Read Only) RBR	0 DLAB=0 Transmitter Holding Register (Write Only) THR	1 DLAB=0 Interrupt Enable Register IER	2 Interrupt Ident. Register (Read Only) IIR	3 Line Control Register LCR	4 MODEM Control Register MCR	5 Line Status Register LSR	6 MODEM Status Register MSR	0 DLAB=1 Divisor Latch (LS) DLL	1 DLAB=1 Division Latch (MS) DLM
0	Data Bit 0 (Note 1)	Data Bit 0	Received Data Available	"0" if Interrupt Pending	Word Length Select Bit 0 (WLS0)	Data Terminal Ready (DTR)	Data Ready (DR)	Delta 0 Clear to Send (DCTS)	Bit 0	Bit 8
1	Data Bit 1	Data Bit 1	Transmitter Holding Register Empty	Interrupt ID Bit (0)	Word Length Select Bit 1 (WLS1)	Request to Send (RTS)	Overrun Error (OE)	Delta Data Set Ready (DDSR)	Bit 1	Bit 9
2	Data Bit 2	Data Bit 2	Receiver Line Status	Interrupt ID Bit (1)	Number of Stop Bits (STB)	Out 1	Parity Error (PE)	Trailing Edge Ring Indicator (TERI)	Bit 2	Bit 10
3	Data Bit 3	Data Bit 3	MODEM Status	0	Parity Enable (PEN)	Out 2	Framing Error (FE)	Delta Data Carrier Detect (DDCD)	Bit 3	Bit 11
4	Data Bit 4	Data Bit 4	0	0	Even Parity Select (EPS)	Loop	Break Interrupt (BI)	Clear to Send (CTS)	Bit 4	Bit 12
5	Data Bit 5	Data Bit 5	0	0	Stick Parity	0	Transmitter Holding Register (THRE)	Data Set Ready (DSR)	Bit 5	Bit 13
6	Data Bit 6	Data Bit 6	0	0	Set Break	0	Transmitter Shift Register Empty (TSRE)	Ring Indicator (RI)	Bit 6	Bit 14
7	Data Bit 7	Data Bit 7	0	0	Divisor Latch Access Bit (DLAB)	0	0	Data Carrier Detect (DCD)	Bit 7	Bit 15

Note 1: Bit 0 is the least significant bit. It is the first bit serially transmitted or received.

8.0 Registers (Continued)

and 1, one and a half Stop bits are generated. If bit 2 is a logic 1 when either a 6-, 7-, or 8-bit word length is selected, two Stop bits are generated. The Receiver checks the first Stop bit only, regardless of the number of Stop bits selected.

Bit 3: This bit is the Parity Enable bit. When bit 3 is a logic 1, a Parity bit is generated (transmit data) or checked (receive data) between the last data word bit and Stop bit of the serial data. (The Parity bit is used to produce an even or odd number of 1s when the data word bits and the Parity bit are summed.)

Bit 4: This bit is the Even Parity Select bit. When bit 3 is a logic 1 and bit 4 is a logic 0, an odd number of logic 1s is transmitted or checked in the data word bits and Parity bit. When bit 3 is a logic 1 and bit 4 is a logic 1, an even number of logic 1s is transmitted or checked.

Bit 5: This bit is the Stick Parity bit. When bits 3, 4 and 5 are logic 1 the Parity bit is transmitted and checked as a logic 0. If bits 3 and 5 are 1 and bit 4 is a logic 0 then the Parity bit is transmitted and checked as a logic 1. If bit 5 is a logic 0 Stick Parity is disabled.

Bit 6: This bit is the Break Control bit. It causes a break condition to be transmitted by the UART. When it is set to a logic 1, the serial output (SOUT) is forced to the Spacing (logic 0) state. The break is disabled by clearing bit 6 to a logic 0. The Break Control bit acts only on SOUT and has no effect on the transmitter logic.

Note: This feature enables the CPU to alert a terminal in a computer communications system. If the following sequence is used no erroneous or extraneous characters will be transmitted because of the break.

1. Load an all 0s, pad character, in response to THRE.
2. Set break after the next THRE.
3. Wait for the transmitter to be idle, (TSRE = 1), and clear break when normal transmission has to be restored.

During the break, the Transmitter can be used as a character timer to accurately establish the break duration.

Bit 7: This bit is the Divisor Latch Access Bit (DLAB). It must be set high (logic 1) to access the Divisor Latches of the Baud Generator during a Read or Write operation. It must be set low (logic 0) to access the Receiver Buffer, the Transmitter Holding Register, or the Interrupt Enable Register.

8.2 Typical Clock Circuits

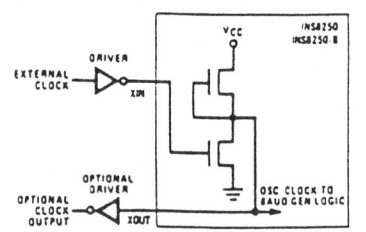

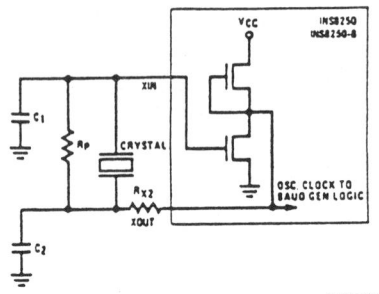

Typical Oscillator Networks

Crystal	R_P	R_{X2}	C_1	C_2
1.8–3.1 MHz	1 MΩ	1.5k	10–30 pF	40–60 pF

TABLE III. Baud Rates Using 1.8432 MHz Crystal

Desired Baud Rate	Decimal Divisor Used to Generate 16 x Clock	Percent Error Difference Between Desired and Actual
50	2304	—
75	1536	—
110	1047	0.026
134.5	857	0.058
150	768	—
300	384	—
600	192	—
1200	96	—
1800	64	—
2000	58	0.69
2400	48	—
3600	32	—
4800	24	—
7200	16	—
9600	12	—
19200	6	—
38400	3	—
56000	2	2.86

TABLE IV. Baud Rates Using 3.072 MHz Crystal

Desired Baud Rate	Decimal Divisor Used to Generate 16 x Clock	Percent Error Difference Between Desired and Actual
50	3840	—
75	2560	—
110	1745	0.026
134.5	1428	0.034
150	1280	—
300	640	—
600	320	—
1200	160	—
1800	107	0.312
2000	96	—
2400	80	—
3600	53	0.628
4800	40	—
7200	27	1.23
9600	20	—
19200	10	—
38400	5	—

8.0 Registers (Continued)

8.3 PROGRAMMABLE BAUD GENERATOR

The UART contains a programmable Baud Generator that is capable of taking any clock input from DC to 3.1 MHz and dividing it by any divisor from 1 to $2^{16}-1$. The output frequency of the Baud Generator is 16 × the Baud [divisor # = (frequency input) ÷ (baud rate × 16)]. Two 8-bit latches store the divisor in a 16-bit binary format. These Divisor Latches must be loaded during initialization in order to ensure proper operation of the Baud Generator. Upon loading either of the Divisor Latches, a 16-bit Baud counter is immediately loaded.

Tables III and IV provide decimal divisors to use with crystal frequencies of 1.8432 MHz and 3.072 MHz, respectively, for common baud rates. For baud rates of 38400 and below, the error obtained is minimal. The accuracy of the desired baud rate is dependent on the crystal frequency chosen. Using a division of 0 is not recommended.

Note: The maximum operating frequency of the Baud Generator is 3.1 MHz. However, when using divisors of 3 and below, the maximum frequency is equal to the divisor in MHz. For example, if the divisor is 1, then the maximum frequency is 1 MHz. In no case should the data rate be greater than 56k Baud.

8.4 LINE STATUS REGISTER

This 8-bit register provides status information to the CPU concerning the data transfer. Table II shows the contents of the Line Status Register. Details on each bit follow:

Bit 0: This bit is the receiver Data Ready (DR) indicator. Bit 0 is set to a logic 1 whenever a complete incoming character has been received and transferred into the Receiver Buffer Register. Bit 0 is reset to a logic 0 by reading the data in the Receiver Buffer Register.

Bit 1: This bit is the Overrun Error (OE) indicator. Bit 1 indicates that data in the Receiver Buffer Register was not read by the CPU before the next character was transferred into the Receiver Buffer Register, thereby destroying the previous character. The OE indicator is set to a logic 1 upon detection of an overrun condition and reset whenever the CPU reads the contents of the Line Status Register.

Bit 2: This bit is the Parity Error (PE) indicator. Bit 2 indicates that the received data character does not have the correct even or odd parity, as selected by the even-parity-select bit. The PE bit is set to a logic 1 upon detection of a parity error and is reset to a logic 0 whenever the CPU reads the contents of the Line Status Register.

Bit 3: This bit is the Framing Error (FE) indicator. Bit 3 indicates that the received character did not have a valid Stop bit. Bit 3 is set to a logic 1 whenever the Stop bit following the last data bit or parity bit is a logic 0 (Spacing level). The FE indicator is reset whenever the CPU reads the contents of the Line Status Register. The UART will try to resynchronize after a framing error. To do this it assumes that the framing error was due to the next start bit, so it samples this "start" bit twice and then takes in the "data".

Bit 4: This bit is the Break Interrupt (BI) indicator. Bit 4 is set to a logic 1 whenever the received data input is held in the Spacing (logic 0) state for longer than a full word transmission time (that is, the total time of Start bit + data bits + Parity + Stop bits). The BI indicator is reset whenever the CPU reads the contents of the Line Status Register. Restarting after a break is received, requires the SIN pin to be logical 1 for at least ½ bit time.

Note: Bits 1 through 4 are the error conditions that produce a Receiver Line Status interrupt whenever any of the corresponding conditions are detected and the interrupt is enabled.

Bit 5: This bit is the Transmitter Holding Register Empty (THRE) indicator. Bit 5 indicates that the UART is ready to accept a new character for transmission. In addition, this bit causes the UART to issue an interrupt to the CPU when the Transmit Holding Register Empty Interrupt enable is set high. The THRE bit is set to a logic 1 when a character is transferred from the Transmitter Holding Register into the Transmitter Shift Register. The bit is reset to logic 0 whenever the CPU loads the Transmitter Holding Register.

Bit 6: This bit is the Transmitter Shift Register Empty (TSRE) indicator. Bit 6 is set to a logic 1 whenever the Transmitter Shift Register (TSR) is empty. It is reset to a logic 0 whenever a data character is transferred to the TSR.

Bit 7: This bit is permanently set to logic 0.

Note: The Line Status Register is intended for read operations only. Writing to this register is not recommended as this operation is only used for factory testing.

TABLE V. Interrupt Control Functions

Interrupt Identification Register				Interrupt Set and Reset Functions		
Bit 2	Bit 1	Bit 0	Priority Level	Interrupt Type	Interrupt Source	Interrupt Reset Control
0	0	1	—	None	None	—
1	1	0	Highest	Receiver Line Status	Overrun Error or Parity Error or Framing Error or Break Interrupt	Reading the Line Status Register
1	0	0	Second	Received Data Available	Receiver Data Available	Reading the Receiver Buffer Register
0	1	0	Third	Transmitter Holding Register Empty	Transmitter Holding Register Empty	Reading the IIR Register (if source of interrupt) or Writing into the Transmitter Holding Register
0	0	0	Fourth	MODEM Status	Clear to Send or Data Set Ready or Ring Indicator or Data Carrier Detect	Reading the MODEM Status Register

8.0 Registers (Continued)

8.5 INTERRUPT IDENTIFICATION REGISTER

In order to provide minimum software overhead during data character transfers, the UART prioritizes interrupts into four levels and records these in the Interrupt Identification Register. The four levels of interrupt conditions in order of priority are Receiver Line Status; Received Data Ready; Transmitter Holding Register Empty; and MODEM Status.

When the CPU accesses the IIR, the UART freezes all interrupts and indicates the highest priority pending interrupt to the CPU. While this CPU access is occurring, the UART records new interrupts, but does not change its current indication until the access is complete. Table II shows the contents of the IIR. Details on each bit follow:

Bit 0: This bit can be used in an interrupt environment to indicate whether an interrupt condition is pending. When bit 0 is a logic 0, an interrupt is pending and the IIR contents may be used as a pointer to the appropriate interrupt service routine. When bit 0 is a logic 1, no interrupt is pending.

Bits 1 and 2: These two bits of the IIR are used to identify the highest priority interrupt pending as indicated in Table V.

Bits 3 through 7: These five bits of the IIR are always logic 0.

8.6 INTERRUPT ENABLE REGISTER

This register enables the four types of UART interrupts. Each interrupt can individually activate the interrupt (INTR) output signal. It is possible to totally disable the interrupt system by resetting bits 0 through 3 of the Interrupt Enable Register (IER). Similarly, setting bits of this register to a logic 1, enables the selected interrupt(s). Disabling an interrupt prevents it from being indicated as active in the IIR and from activating the INTR output signal. All other system functions operate in their normal manner, including the setting of the Line Status and MODEM Status Registers. Table II shows the contents of the IER. Details on each bit follow.

Bit 0: This bit enables the Received Data Available Interrupt when set to logic 1.

Bit 1: This bit enables the Transmitter Holding Register Empty Interrupt when set to logic 1.

Bit 2: This bit enables the Receiver Line Status Interrupt when set to logic 1.

Bit 3: This bit enables the MODEM Status Interrupt when set to logic 1.

Bits 4 through 7: These four bits are always logic 0.

8.7 MODEM CONTROL REGISTER

This register controls the interface with the MODEM or data set (or a peripheral device emulating a MODEM). The contents of the MODEM Control Register (MCR) are indicated in Table II and are described below. Table II shows the contents of the MCR. Details on each bit follow.

Bit 0: This bit controls the Data Terminal Ready ($\overline{DTR}$) output. When bit 0 is set to a logic 1, the $\overline{DTR}$ output is forced to a logic 0. When bit 0 is reset to a logic 0, the $\overline{DTR}$ output is forced to a logic 1.

Note: The $\overline{DTR}$ output of the UART may be applied to an EIA inverting line driver (such as the DS1488) to obtain the proper polarity input at the succeeding MODEM or data set.

Bit 1: This bit controls the Request to Send ($\overline{RTS}$) output. Bit 1 affects the $\overline{RTS}$ output in a manner identical to that described above for bit 0.

Bit 2: This bit controls the Output 1 ($\overline{OUT\ 1}$) signal, which is an auxiliary user-designated output. Bit 2 affects the $\overline{OUT\ 1}$ output in a manner identical to that described above for bit 0.

Bit 3: This bit controls the Output 2 ($\overline{OUT\ 2}$) signal, which is an auxiliary user-designated output. Bit 3 affects the $\overline{OUT\ 2}$ output in a manner identical to that described above for bit 0.

Bit 4: This bit provides a local loopback feature for diagnostic testing of the UART. When bit 4 is set to logic 1, the following occur: the transmitter Serial Output (SOUT) is set to the Marking (logic 1) state; the receiver Serial Input (SIN) is disconnected; the output of the Transmitter Shift Register is "looped back" into the Receiver Shift Register input; the four MODEM Control inputs ($\overline{CTS}$, $\overline{DSR}$, $\overline{RI}$, and $\overline{DCD}$) are disconnected; and the four MODEM Control outputs ($\overline{DTR}$, $\overline{RTS}$, $\overline{OUT\ 1}$, and $\overline{OUT\ 2}$) are internally connected to the four MODEM Control inputs. In the diagnostic mode, data that is transmitted is immediately received. This feature allows the processor to verify the transmit-and received-data paths of the UART.

In the diagnostic mode, the receiver and transmitter interrupts are fully operational. The MODEM Control Interrupts are also operational, but the interrupts' sources are now the lower four bits of the MODEM Control Register instead of the four MODEM Control inputs. The interrupts are still controlled by the Interrupt Enable Register.

Bits 5 through 7: These bits are permanently set to logic 0.

8.8 MODEM STATUS REGISTER

This register provides the current state of the control lines from the MODEM (or peripheral device) to the CPU. In addition to this current-state information, four bits of the MODEM Status Register provide change information. These bits are set to a logic 1 whenever a control input from the MODEM changes state. They are reset to logic 0 whenever the CPU reads the MODEM Status Register.

Table II shows the contents of the MSR. Details on each bit follow:

Bit 0: This bit is the Delta Clear to Send (DCTS) indicator. Bit 0 indicates that the $\overline{CTS}$ input to the chip has changed state since the last time it was read by the CPU.

Bit 1: This bit is the Delta Data Set Ready (DDSR) indicator. Bit 1 indicates that the $\overline{DSR}$ input to the chip has changed state since the last time it was read by the CPU.

Bit 2: This bit is the Trailing Edge of Ring Indicator (TERI) detector. Bit 2 indicates that the $\overline{RI}$ input to the chip has changed from a low to a high state.

Bit 3: This bit is the Delta Data Carrier Detect (DDCD) indicator. Bit 3 indicates that the $\overline{DCD}$ input to the chip has changed state.

Note: Whenever bit 0, 1, 2, or 3 is set to logic 1, a MODEM Status Interrupt is generated.

Bit 4: This bit is the complement of the Clear to Send ($\overline{CTS}$) input. If bit 4 (loop) of the MCR is set to a 1, this bit is equivalent to RTS in the MCR.

Bit 5: This bit is the complement of the Data Set Ready ($\overline{DSR}$) input. If bit 4 of the MCR is set to a 1, this bit is equivalent to DTR in the MCR.

Bit 6: This bit is the complement of the Ring Indicator ($\overline{RI}$) input. If bit 4 of the MCR is set to a 1, this bit is equivalent to OUT 1 in the MCR.

Bit 7: This bit is the complement of the Data Carrier Detect ($\overline{DCD}$) input. If bit 4 of the MCR is set to a 1, this bit is equivalent to OUT 2 in the MCR.